Environmental Science and Engineering

Environmental Science

For further volumes:
http://www.springer.com/series/3234

Hiroshi Kitazato • Joan M. Bernhard
Editors

Approaches to Study Living Foraminifera

Collection, Maintenance and Experimentation

Springer

Editors
Hiroshi Kitazato
Japan Agency for Marine-Earth Science and Technology
Kanagawa, Japan

Joan M. Bernhard
Woods Hole Oceanographic Institution
Massachusetts, USA

ISSN 1863-5520 ISSN 1863-5539 (electronic)
ISBN 978-4-431-54387-9 ISBN 978-4-431-54388-6 (eBook)
DOI 10.1007/978-4-431-54388-6
Springer Tokyo Heidelberg New York Dordrecht London

Library of Congress Control Number: 2013955595

Printed on acid-free paper

Springer is part of Springer Science+Business Media (www.springer.com)

Preface

Foraminifera are eukaryotic micro-organisms that are mainly distributed in oceanic environments. This protistan taxon plays a major role in nutrient cycling of the oceanic hydrosphere and in biogeochemical cycling. Because these microbes sometimes compose more than 50 % of the biomass in the deep oceans as well as in some high-latitude settings, they are a key taxon in marine ecosystems. Foraminifera inhabit both the benthic and the pelagic realms. The skeletal component of foraminifera are referred to as tests (i.e., shells), which can be composed of calcium carbonate, agglutinated particles, or organic (non-mineralized) material secreted by the foraminifer.

Because each foraminifer makes at least one test during its life cycle, foraminiferal tests are commonly found in marine sediments. The majority of these tests can become preserved in marine sedimentary deposits. These shells are major components of marine strata. Fossil foraminifera, thus, have long been used, since the dawn of geological investigations. Foraminiferal index fossils are useful for both assigning geologic ages and reconstructing paleoenvironments of strata. In this context, an enormous amount of data has been accumulated in terms of modern foraminiferal distribution in the oceans and stratigraphic records of foraminifera during the Phanerozoic.

Data suggest that foraminifera may show rapid rates of morphological evolution through geologic time. From studies of modern environments, it also can be asserted that foraminifera are quite sensitive to environmental changes. Thus, interpretations of the fossil record can lend insights into paleobiology. However, we often lack the knowledge of living foraminiferal species to clearly understand the paleobiology of extinct species.

During the early history of foraminiferal research, in the late nineteenth and early twentieth centuries, the natural history of foraminifera was well studied by naturalists from North America, South America, and Europe. After foraminifera were established as useful indicators for age determinations and paleoenvironmental reconstruction of strata, research on foraminifera shifted to method development for both precise biochronology and paleoceanographic proxies using elemental compositions in their tests. Recently, isotopic and trace elements of foraminiferal tests are major tools used to interpret paleoceanographic records.

While paleontological and paleoceanographic proxies of foraminifera were being developed, basic biological research on foraminifera was being conducted by biologists such as J. Le Calvez, Ernst Meyers, Esteban Boltovskoy, J. S. Bradshaw, Karl G. Grell, Zach Arnold, Alan Bé, and John J. Lee. They discovered and applied fundamental knowledge about the natural history of foraminifera. For instance, they established the life cycles, basic cell anatomy and physiology, food preferences, ecological niches, and other details on available species. They also developed culturing and observational techniques for working on living foraminifera.

During these recent decades, particularly during the twenty-first century, biological research on foraminifera has been expanding in the number of scientists involved as well as in the diversity of taxonomic groups of foraminifera being studied. Both laboratory and in situ experimental research are being conducted at laboratories worldwide. Some of these studies are being aided by the development of sophisticated culture systems. On the other hand, basic knowledge of the biological approaches has not been extensively shared among scientists who are working on living foraminifera. Sharing basic experiences among scientists from many different nationalities is critical in the development of cutting-edge collaborations to make progress in our science on living foraminifera. Technology and experience transfer from talented scientists to beginning investigators is also an important and effective means to propagate biological research of foraminifera on a global scale.

With these backgrounds and concepts in mind, in July 2012, we carried out a field workshop by convening scientists who investigate aspects of living foraminifera. The workshop was held in Japan, where diverse marine environments, from subtropical to subarctic oceanic realms and from intertidal/littoral to ultra-hadal depth ranges, exist around the Japanese Islands. Forty-five scientists from 13 countries and regions gathered in one place, collected, maintained, and observed living foraminifera, and discussed various research topics related to living planktic and benthic foraminifera for 1 week. We shared many facts and basic techniques about the biology of foraminifera through the workshop. In particular, we formed strong functional research networks among all attendees from around the world. The agenda of the workshop is included with this preface.

Field Workshop on Living Foraminifera in Japan

July 15th (Sun.) ~ 21 st (Sat.), 2012

Program

Sunday, 15th July, 2012

15:00 ~ 20:00		Meet at Daiwa Roynet Hotel, Kokusaidori in Naha City, Okinawa, Registration
18:00 ~ 20:00		Opening Ice Breaker

Monday, 16th July, 2012

7:30 ~		Transfer from Naha to Sesoko Marine Biological Station (Depart at 8:30 AM, Daiwa Roynet Hotel)
9:30 ~		Okinawa Churaumi Aquarium & backyard tour
12:00 ~ 13:30		**Lunch @ Aquarium**
14:30		Guidance (Opening: Kitazato, guidance: Sesoko Marine Station) @ Marine Resort Bellvue Hotel
		Introduction: Sampling, observatioon and culture: Moderator (Hiroshi Kitazaro & Johann Hohenegger)
		Self-introduction (3 min./person)
18:00 ~		Dinner @ Marine Resort Bellvue Hotel

Tuesday, 17th July, 2012

7:00 ~		Breakfast @ Marine Resort Bellvue Hotel
8:00 ~ 8:30		Transfer from Hotel to Sesoko Marine Station
9:00 ~ 14:30		Sampling & culturing (with boat or at the beach)
(12:00 ~ 13:30)		Lunch @ Sesoko (Lunch box)
14:30 ~ 15:00		Transfer from Sesoko to Hotel
15:00 ~ 15:30	Joan M. Bernhard	Were foraminifera the beginning of the end for Stromatolites?
15:30 ~ 16:00		***Larger benthic foraminifer & host-symbiont interaction: Moderator (Kazuhiko Fujita & Johann Hohenegger)***
16:00 ~ 16:15	Christiane Schmidt	Effects of global change stress on symbiont-bearing benthic Foraminifera from the Great Barrier Reef
16:15 ~ 16:30	Benjamin James Ross	Developing protocols for testing toxic chemicals on the symbiotic foraminifer, *Amphistegina gibbosa* d'Orbigny: Challenges and new directions
16:30 ~ 16:45		**Coffee Break**
16:45 ~ 17:00	Ahuva Almogi-Labin	*Pararotalia spinigera* (Le Calvez) alien symbiont-bearing foraminifera flourishing on Israeli rocky coasts: Field and culture records
17:00 ~ 17:15	Laura Jane Cotton	Extinction of larger benthic foraminifera at the Eocene/Oligocene boundary
17:15 ~ 17:30	Antonino Briguglio	Ontogenetic study of living and fossil nummulitids by micro-computed tomography
17:30 ~ 18:00		***Ocean acidification & culture experiment: Moderator (Katsunori Kimoto)***
18:00 ~ 18:15	Sutinee Sinutok	Warmer more acidic oceans reduce productivity and calcification in the symbiont-bearing foraminifera *Marginopora vertebralis*
18:15 ~ 18:30	Mana Hikami	Effects of high pCO_2 seawater on symbiont-bearing reef foraminifers
19:00 ~		Dinner @ Marine Resort Bellvue Hotel

Wednesday, 18th July, 2012

7:00 ~		Breakfast @ Marine Resort Bellvue Hotel
8:00 ~ 8:30		Transfer from Hotel to Sesoko Marine Station
9:00 ~ 14:30		Sampling & culturing (with boat or at the beach)
(12:00 ~ 13:30)		**Lunch @ Sesoko (Lunch box)**
13:30 ~ 14:00		Lab.tour (Harii, Sinniger, Hikami, FUjita & Sakai)
15:00 ~ 15:30		***Planktonic foraminiferal culture experiment : Moderator (Haward J. Spero & Azumi Kuroyanagi)***
15:30 ~ 16:00	Howard J. Spero	Novel Techniques Unravel Calcification Mechanisms and Shell Geochemistry in Planktonic Foraminifera
16:00 ~ 16:15	Azumi Kuroyanagi	Foraminiferal biological response to low dissolved oxygen condition
16:15 16:30	Katsunori Kimoto	Evaluation of biological impacts for marine calcareous zooplankton to ocean acidification using Micro-focus X-ray CT scanning technology
16:30 ~ 16:45		**Coffee Break**
16:45 ~ 17:15		***Molecular & environmental assesment: Moderator (Beatrice Lecroq & Atsushi Kurasawa, Frederic Sinniger)***
17:15 ~ 17:30	Steve Shao-Jen Doo	Assessing resiliency on multiple scales: Using molecular to remote sensing data to synthesize effects of climate change on large benthic Foraminifera communities.
17:30 ~ 17:45	Gily Merkado	Breaking biogeographic barriers: Molecular and morphological evidences for the Lessepsian invasion of soritid foraminifers to the Mediterranean Sea
18:00 ~ 20:00		**BBQ@Sesoko Marine Station**
20:00		Transfer from Sesoko to Hotel

Thursday, 19th July, 2012

8:00 ~ 11:00		Transfer from Sesoko Marine Biological Station to Naha via GODAC
9:00 ~ 10:30		GODAC, Glopbal Oceanographic Data Center, JAMSTEC in Nago
10:30 ~ 12:00		to Naha Airport
		Naha~Haneda~Kawasaki

Friday, 20th July, 2012

9:00 ~		Meet at Oppama station
9:30 ~ 10:00		***in situ-, in vitro- experiment & population dynamics: Moderator (Joan Bernhard & Hidetaka Nomaki)***
10:00 ~ 10:15	Petra Heinz	Response of shallow-water foraminifera to a C^{13} - labelled food pulse in the laboratory
10:15 ~ 10:30	Shai Oron	Benthic formaminiferal response to the removal of aquaculture fishcages in the Gulf of Aqaba-Eilat
10:30 ~ 10:45	Pauline Duros	Ecology of benthic foraminifera in the Whittard canyon (Bay of Biscay, North Atlantic): distribution and composition of foraminiferal assemblages, geochemistry of sediments
10:45 ~ 11:00		**Coffee Break**
11:00 ~ 11:30		***Environmetal resposes: Moderator (Emmanuelle Geslin)***
11:30 ~ 12:00	Emmanuelle Geslin	Benthic foraminifer responses to abiotic and biotic environmental forcing
12:00 ~ 13:30		**Lunch @ JAMSTEC (Lunch box)**

Friday, 20th July, 2012 (PM)

13:30 ~ 14:30	***Lab.tour: culture systems (Hidetaka Nomaki, Takashi Toyofuku & Katsunori Kimoto)***	
14:30 ~ 14:45	**Coffee Break**	
	(Continued) Environmetal resposes: Moderator (Emmanuelle Geslin)	
14:45 ~ 15:00	Andre Rosch Rodrigues	Living foraminifera in Brazilian subtropical coastal environment (Flamengo Inlet, Ubatuba, SP)
15:00 ~ 15:15	Christine Barras	Effect of induced hypoxia on benthic foraminiferal mortality, reproduction and growth
15:15 ~ 15:30	Manuel Fritz Gerhard Weinkaut	Morphological response of Foraminifera to environmental stress during the deposition of a Mediterranean sapropel
15:30 ~ 15:45	**Coffee Break**	
15:45 ~ 16:15	***Molecular phylogeny: Moderator (Yurika Ujiie & Masashi Tsuchiya)***	
16:15 ~ 16:30	Yurika Ujiie	Longitudinal differentiation among pelagic populations in a planktic foraminifer
16:30 ~ 16:45	Atsushi Kurasawa	Genetic diversity of *Globigerina bulloides* in the Pacific Ocean
16:45 ~ 17:00	Beatrice Lecroq	Ultra-deep sequencing of ultra-deep sediment
17:30 ~	to Oppama Stn.	

Saturday, 21st July, 2012

8:30 ~	Meet at Oppama station~transfer to Enoshima Is.	
9:30 ~ 10:00	***Calcification: Moderator (Lennart de Nooijer & Takashi Toyofuku)***	
10:00 ~ 10:30	Lennart de Nooijer	Minor and Trace Element Incorporation in *Ammonia tepida*
10:30 ~ 10:45	Kazuhiko Ichikawa	Cellular Calcification and Physiology of Planktonic Foraminifera
10:45 ~ 11:00	Annette Bolton	Examination of trace element variability in fossil planktonic foraminifera using laser ablation inductively coupled plasma mass spectrometry (LA-ICPMS)
11:00 ~ 11:15	Helena L. Filipsson	*Bulimina aculeata/marginata* cultured over a large temperature gradient: d^{18}O and Mg/Ca results
11:15 ~ 11:30	Stephen Eggins	Has the sensitivity of Mg/Ca seawater paleothermometers always been the same?
11:30 ~ 11:45	Kate Holland	How do planktic foraminifers control the Mg composition of their tests?
11:45 ~ 12:00	***Wrap up session & closing remarks: Hiroshi Kitazato***	
12:00 ~ 13:00	**Lunch @ Enoshima Is. (Lunch box)**	
13:30 ~ 16:30	visit New Enoshima Aquarium	
18:00 ~ 20:00	WS dinner & Workshop closing ceremony	
20:00 ~	to Kawasaki	

Because many researchers who wish to study living foraminifera typically begin with a geological and/or paleontological background rather than a biological background, it can be difficult for some to begin working on living foraminifera. Also, due to foraminiferal small size and slow movements, it can be difficult to know how and where foraminifera are living, how they can be collected, and how they can be maintained in a healthy manner in the laboratory.

On the occasion of this workshop, we planned to publish a book to facilitate sharing of basic knowledge about how to conduct research on living foraminifera in the field and in the laboratory. This book is basically composed of papers by participants in the field workshop, although some papers were invited to augment particular aspects of the field. Through this book, we attempt to show our knowledge about living foraminifera and research techniques and applications relevant, and sometimes unique, to foraminifera. Basic topics and research methodologies are presented in this book: for instance, collection of living foraminifera in the field, maintenance of foraminifera in the laboratory, the conceptual framework of experiments, and various analytical methods including molecular analyses. The reader is, of course, directed to the primary literature for many methods too numerous to cover in this contribution. Excellent compilations dedicated to living foraminifera include *Biology of Foraminifera* (John J. Lee and O. Roger Anderson; Academic Press, 1991), *Modern Foraminifera* (edited by Barun K. Sen Gupta; Kluwer, 1999), and *Advances in the Biology of Foraminifera* (a special issue of *Micropaleontology* edited by John J. Lee and Pamela Hallock; 2000). Our book should serve as a guidebook to introduce scientists into the world of living foraminifera as well as provide details on methodological developments of the past decade.

We believe that foraminifera could serve as a model organism for life and environmental sciences. With an evolutionary history dating back to at least the early Cambrian, foraminifera have had ample time to explore every available niche. They are known to have high mutation rates and, as noted, to react sensitively to environmental changes. Furthermore, foraminifera are able to inhabit nearly every marine habitat (and freshwater as well as terrestrial environs), including some extreme habitats such as anoxic and sulfide-enriched milieus. However, to establish any taxon as a model organism for biological study, we should fully understand its fundamental biology as well as natural history. In the case of foraminifera, we need to gather much greater comprehension of their basic cell biology, ecology, and evolution. For example, at the time of this printing, not one total genome of any foraminiferal species is published. Once such information is available, we will then be able to explore cutting-edge life science questions using their functional gene sequences.

Both Hiroshi Kitazato and Joan M. Bernhard played roles in producing the workshop and this book. Hiroshi Kitazato, JAMSTEC, coordinated the field workshop, envisioned and planned the book, negotiated with Springer Japan, and made final editing decisions. Joan M. Bernhard, WHOI, gave the plenary lecture at the workshop and edited all manuscripts for both scientific and linguistic aspects.

We gratefully acknowledge the following people who played important roles in ensuring the success of the field workshop and production of the book: Kazuhiko Fujita, Kazuhiko Sakai, Saki Sinniger-Harii, Frederic Sinniger-Harii, Ayako Yasumoto, Masashi Tsuchiya, Takashi Toyofuku, Hidetaka Nomaki, Katsunori Kimoto, Yurika Ujiie, Beatrice Lecroq, Pauline Duros, Atsushi Kurasawa, and members of both the University of the Ryukyus and the Institute of Biogeosciences, JAMSTEC. Financial support for the workshop was provided by research institutions and oil companies, the University of the Ryukyus, Tohoku University, JAMSTEC, INPEX Cooperation, JAPEX Co. Ltd., and JX Inc. Kind acknowledgment is also extended to Ken Kimlicka and Taeko Sato for providing the opportunity to publish this book through Springer Japan. J.M.B. also gratefully acknowledges partial financial support from The Robert W. Morse Chair for Excellence in Oceanography from Woods Hole Oceanographic Institution.

We happily publish this book for all people, researchers and enthusiasts alike, wishing to investigate the fascinating world of living foraminifera. Let's play with living foraminifera!

Tokyo, Japan — Hiroshi Kitazato
August 13, 2013
Woods Hole, Massachusetts, USA — Joan M. Bernhard
August 16, 2013

Group photograph in Okinawa, Japan

Group photograph at Yokosuka, Japan

Contents

Chapter 1
The Propagule Method as an Experimental Tool in Foraminiferal Ecology

Elisabeth Alve and Susan T. Goldstein

Abstract The Propagule Method provides a novel and logistically simple experimental tool for examining the ecology of benthic foraminifera, including their dispersal and the responses of both individual species and multi-species assemblages to one or more specific environmental conditions. Propagules, small juveniles, form a "bank" in fine-grained marine sediments following dispersal and settlement. As shown previously, propagules may be derived from local or distant populations, and those of some species may persist in a dormant or cryptic state for months or even years. In general, the Propagule Method first involves separating the propagule bank from mature foraminifera and various larger organisms, which is easily done by sieving. The fine sediment fraction contains the propagule bank and is retained for experimental treatments and associated controls. At the conclusion of the experiment, mature assemblages are harvested following growth and in some cases reproduction. By focusing on the responses of juveniles (propagules) as they grow to adults, this approach is consistent with studies that use invertebrate larvae in that it uses juveniles to assess responses to environmental conditions. The method has the potential to work for a wide range of benthic foraminifera, irrespective of test composition.

Keywords Exotic species • Experimental ecology • Microcosms • Propagule bank

E. Alve (✉)
Department of Geosciences, University of Oslo, P.O. Box 1047, Blindern, 0316 Oslo, Norway
e-mail: ealve@geo.uio.no

S.T. Goldstein
Department of Geology, University of Georgia, Athens, GA 30602, USA

H. Kitazato and J.M. Bernhard (eds.), *Approaches to Study Living Foraminifera: Collection, Maintenance and Experimentation*, Environmental Science and Engineering,
DOI 10.1007/978-4-431-54388-6_1,

1.1 Introduction

Though field-based studies underpin much of our current understanding of foraminiferal ecology (e.g., syntheses by Murray 1991, 2006), a variety of laboratory experiments have been used in recent years to address questions not easily studied in the field. Mesocosms and microcosms containing foraminifera-bearing sediment, for example, have been widely used to examine the responses of assemblages or specific taxa to oxygen availability (e.g., Alve and Bernhard 1995), food input (Ernst and van der Zwaan 2004; Nomaki et al. 2005; Topping et al. 2006; Mojtahid et al. 2011), or the effects of selected pollutants (e.g., Gustafsson et al. 2000; Ernst et al. 2006). Further, microcosms adapted with a calibrated flow-through seawater system have been used to examine the effects of specific parameters on foraminiferal shell chemistry (e.g., Chandler et al. 1996; Wilson-Finelli et al. 1998; de Nooijer et al. 2007; Dissard et al. 2010). The common thread running through these experimental approaches is that they begin with either unprocessed, freshly collected sediment with an assemblage of resident foraminifera (both living and dead), or with isolated, largely mature foraminifera that were selected from sieved or otherwise processed sediment. At the level of assemblages, responses are measured by movement or migration patterns within sediment as foraminifera access oxygen or available food by changes in assemblage composition. The challenge is to consistently distinguish living from dead foraminifera (e.g., Bernhard 2000; Bernhard et al. 2004; Murray 2006, p. 11) throughout the experiment so that responses reflect the activity of living individuals rather than the random distribution of empty shells in the sediment. Alternatively in other experimental approaches, living foraminifera are harvested individually from sediments after treatment (e.g., feeding experiments) to examine responses (e.g., Topping et al. 2006), and in shell-chemistry experiments, individuals of a target species are cultured through multiple generations (Hintz et al. 2006a) or grown from immature specimens (e.g., Hintz et al. 2006b) to assess the response.

The term propagule was first applied to benthic foraminifera by Alve and Goldstein (2002) to refer to "small juveniles, perhaps just the proloculus" (i.e., the initial chamber) that may enter a resting stage following reproduction. The juveniles are passively dispersed either locally or more widely via, e.g., water currents. The "Propagule Method" (Goldstein and Alve 2011), outlined here, is a novel, easily-applied experimental tool in foraminiferal ecology. Foraminiferal propagules are tiny juveniles, often measuring just 10s of microns, that are produced by either sexual or asexual reproduction. Most foraminiferal species in which gametogenesis has been observed are gametogamous (i.e., biflagellated gametes released into seawater) (Goldstein 1997), and the resulting zygotes (sexually produced propagules) appear to disperse extensively, both within and beyond the distribution of their conspecific adults. Further, they form a "propagule bank" in sediments comprising propagules derived locally plus those transported into the area from more distant sources (Alve and Goldstein 2002, 2003). This dispersal strategy is broadly analogous to larval dispersal in many marine invertebrates, except that invertebrate larvae do not form "banks" in sediments, and foraminiferal propagules do not exhibit the complex recruitment behaviors (e.g., Young 1995) so characteristic of invertebrate larvae.

The "Propagule Method" was developed based on the observation that species not present as growth stages or adults at a sampling site nonetheless grew from the fine-grained sediments at that site when exposed to appropriate conditions. Their sudden "appearance" showed that they had been present as small juveniles and that they were able to delay growth until the conditions became appropriate. This ability to delay growth (months to a couple of years for some species; Alve and Goldstein 2010) until presented with appropriate conditions can be used in experimental strategies to examine the responses in foraminiferal species as well as assemblages to changes in selected environmental parameters. Here, the response is measured by the ability of propagules to grow under the experimental conditions.

1.2 Experimental Design

1.2.1 General Outline

There are numerous ways to design propagule experiments but the key elements are to separate the numerous propagules present in surface sediments at the sampling site from the more mature foraminiferal growth stages and larger metazoans by sieving, and to promote growth of species present in the finer size-fraction under controlled environmental conditions (Table 1.1). Where, how and how much sediment to collect depends on the aims of the study. Efforts should be made to keep the sediment close to ambient conditions (particularly temperature) until it is sieved (see Sect. 1.2.2). If relevant, quantitative field samples may be collected (e.g., with a multicorer) and examined for comparative purposes.

Sieving should take place as soon as possible after collection in order to avoid unwanted growth before the experiment commences, using ambient seawater from the collection site, filtered or artificial seawater (Instant Ocean®). Re-sieving the sediment may be advisable, particularly if an interval of more than just a few days has passed between collection and start of the experiment. Re-sieving will remove any individuals that may have grown after collection or mature foraminifera that were accidentally introduced as contaminants during the initial sieving. After sieving, the coarse fraction is fixed or preserved for later examination to assess the in situ living and death assemblages at the time of collection. The fine fraction is left to settle overnight under appropriate environmental conditions, the resulting overlying, clear seawater is then siphoned off and may be used later in the microcosms (previously termed containers or growth chambers by us), the fine fraction is gently homogenized, and aliquots of a fixed volume of (e.g., 12, 15, or 20 mL) are extracted with a large pipette and placed in a series of small, translucent microcosms (e.g., round, 118 mL polypropylene or similar plastic containers) with tightly fitting lids, along with a fixed amount (e.g., 40 mL) of appropriate natural or artificial seawater. Using plastic containers rather than impermeable glass allows some gas diffusion in order to prevent the microcosms from turning anoxic. If needed, the microcosms are further sealed with parafilm to avoid evaporation and maintained in

Table 1.1 Summary of the Propagule Method

Key elements
Isolate propagules present in surface sediments at the sampling site from the more mature foraminiferal growth stages and larger metazoans by sieving
Sediment size-fraction used must be sufficiently large to include proloculi of target species (and generations)
Promote growth of species present in the finer size-fraction under controlled environmental conditions
Experiment design
Collect surface sediment (using a box corer or a multicorer if from submarine locations) and keep at ambient conditions (e.g., temperature) until it is sieved
Collect ambient seawater or use artificial seawater (Instant Ocean®) for sieving the sediment
Allow the fine fraction to settle in ambient or artificial seawater overnight; then siphon off the seawater which may be retained for further use in the experiments; homogenize the sediment; divide into aliquots and place each in a microcosm (t=0)
Unless required by the aim of the study (e.g., feeding), microcosms are kept sealed during the course of the experiment
Aliquots from the same source can be subjected to different treatments (with replicates) for comparative purposes, while some are kept as controls
At the end (t=1) promptly sieve the microcosm sediment on a screen having a slightly larger aperture than that used to set up the experiment (to eliminate the sediment and the original foraminiferal tests); all the remaining foraminiferal tests (whether live or dead at t=1) must be the result of growth and/or reproduction

a temperature (and possibly humidity-) regulated climate chamber which may be illuminated with artificial, broad-spectrum lighting set to an appropriate illumination cycle (e.g., 12 h) to promote algal growth in the microcosms or kept in the dark to mimic sub-photic zone conditions. The number of controls and replicates are adjusted according to the questions asked (see also Sect. 1.2.4). To further ensure a clean separation of the propagule bank from mature foraminifera in the original sediment, a set of additional replicates are harvested at t=0 and examined.

All microcosms remain sealed throughout the treatment period unless e.g., feeding is part of the experiment. Such closed systems, static microcosms, with no circulation of sea-water between microcosms prevent dispersal of propagules within the circulation system (e.g., abundant *Rosalina vilardeboana* recorded in circulating laboratory setups, Hintz et al. 2004). As opposed to experiments that either do not use any sediment at all (e.g., Toyofuku and Kitazato 2005) or use artificial sediment (e.g., Hintz et al. 2004) or a specific size-fraction of the ambient, natural sediment (e.g., de Nooijer et al. 2007) as a substrate in the microcosms, the Propagule Method uses the complete, ambient, fine-grained sediment.

1.2.2 Size Fraction of Sediment Used

The choice of sediment size fraction used in a particular experiment depends on the aim of the study and on the size of the proloculus (first-formed chamber) of the target species. Most benthic foraminiferal life cycles include alternations between sexually produced (diploid) and asexually produced (haploid) generations (for review, see Goldstein 1999 and references therein). The generations are commonly distinguished by shell dimorphism, i.e., the size of the proloculus may vary between generations to form small microspheric (sexually formed) and larger megalospheric (asexually formed) proloculi. Hence, the size of the proloculus may not only vary between, but also within, single species. This implies that the upper size limit of the sediment used in the experiment should allow inclusion of the megalospheric forms of the target species. Information on prolocular sizes in different species is not readily available. Exceptions are Höglund (1947), which includes detailed information on characteristics of species in the Skagerrak area (eastern North Sea) and Hofker (1951), who detailed prolocular diameters for selected species from the Siboga Expedition (Indonesia) and portions of the Mediterranean. Until now, we have used sediments <32, < 53 or <63 μm in size to grow shallow-water to bathyal species but if the goal is to grow species such as *Liebusella goësi* or *Globobulimina turgida* with megalospheric prolocular diameters of 90–190 and 80–143μm, respectively (Höglund 1947), a larger size fraction of sediment should be used.

1.2.3 Duration of Experiments

The duration of propagule experiments depends on the growth and reproduction rates of the species present in the propagule bank. Although it is commonly stated that benthic foraminifera have short life cycles (days to months) accurate observations documenting this are few and based on laboratory observations (e.g., Myers 1935; Grell 1956; Pawlowski and Lee 1992). In field studies, conclusions can only be inferred from time-series (which in turn are influenced by frequency of e.g., sampling, patchiness) because, for continuously reproducing species (i.e., some young always present), it is not possible to follow the growth of cohorts (discussion in Murray and Alve 2000).

Our experience so far is that for intertidal assemblages grown with illumination, six weeks is sufficient to produce abundant populations of all growth stages of species for which the given conditions are appropriate. For example, growth and reproduction over just 6 weeks yielded from 323 to 1,325 individuals from initial aliquots of 20 ml of fine sediment (Goldstein and Alve 2011). For these assemblages, increasing the duration of the experiment to 12 weeks does not alter the outcome to any appreciable degree (Goldstein and Alve 2011). Symbiosis with microalgae is known to promote growth in larger foraminifera (e.g., Röttger 1976; Hallock 1981) and it is possible that either chloroplast sequestration (e.g., *Haynesina germanica*, Lopez 1979) or feeding on freshly produced algae under the influence of daylight has a similar positive effect on growth. On the other hand, there are indications that, when kept in the dark, species that harbor symbiotic microalgae grow more slowly (Duguay 1983) and deep-sea species are known to survive in the laboratory for several years without reproducing (Hemleben and Kitazato 1995). Consequently, longer exposure times (e.g., several months) may be needed for some species adapted to life below the photic zone.

1.2.4 Harvesting

At the conclusion of the experiment (t = 1) the microcosms consist of the fine-grained sediment from the start of the experiment (t=0), as well as living individuals and dead (empty) tests with a size larger than the surrounding, fine sediments. The living individuals may have grown throughout, or only at the end of the experimental period, or they are the result of reproduction during this period. As reproduction commonly causes the life of the parent to end (cell content used to produce gametes or juveniles; e.g., Lister 1895), the presence of empty tests will reflect growth and subsequent reproduction and/or death of the organism during the experimental period. However, since the parent tests in some species are destroyed during reproduction (e.g., asexual reproduction in *Trochammina hadai*, Kitazato 1988), reproduction may have occurred even if no dead tests remain. Consequently, the living individuals in the microcosms are those that live at t = 1, and the empty tests represent the minimum number of individuals that have either reproduced or died. On the other hand, the "entire assemblage" includes all tests (whether live or dead) which have responded positively to the controlled experimental conditions by growing during a known time-interval. These differ from "total assemblages" (used in some field studies, discussion in Murray 2000), which include both living individuals and a blend of allochtonous and autochtonous empty tests drawn from an unknown mixture of microhabitats averaged over an unknown interval of time and seasons.

Based on the above, the sediment in each microcosm at t=1 is either processed immediately or, depending on the aim of the study, preserved or fixed to distinguish between live and dead individuals. If, for example, the aim is to test for the occurrence of exotic or allochthonous species in the propagule bank at a site (i.e., propagules transported to a site where conspecific adults do not live), all individuals that have grown by t=1 (i.e., the entire assemblage) should be considered. If the aim also includes obtaining information on reproduction and generation time, separate data on live and dead populations are needed.

The entire assemblages are harvested at t = 1 by immediately sieving the contents of each microcosm separately or fixing/preserving them. The sieve used for harvesting should have larger openings than that used initially (t=0) to process the sediment from the collection site. For example, Goldstein and Alve (2011) used a 53-μm stainless-steel sieve at t=0, and a 63-μm sieve to harvest the assemblages at t=1. This ensures that those foraminifera harvested had grown and therefore responded positively to the experimental conditions. Additionally, it removes the sediment and thereby simplifies examination of the assemblages. Hence, this experimental design allows for relatively quick assessment of assemblages and should encourage the inclusion of sufficient replicates and controls.

1.3 Strengths and Weaknesses of the Propagule Method

In ecology, there is a distinction between survival and growth. The latter requires more favorable conditions than the former. Consequently, if the aim of a study is not only to investigate whether certain conditions are acceptable for survival but also for growth and possibly reproduction, the method used needs to allow differentiation between individuals grown before and after the onset of the experiment. The Propagule Method allows just this because it is indisputable that all populations present in the microcosms at the end of experiments have responded positively (i.e., by growth/reproduction) to the treatment (assuming no contamination, see Sect. 1.2.1 above). This approach also has the advantage of focusing on the critical, juvenile developmental stages, i.e., using juveniles rather than adults as the former are generally considered to be more sensitive to environmental pressure (e.g., Olsgard 1999; Pineda et al. 2012). Additionally, the method does not require a lot of time-consuming work isolating and picking individuals, assumed to be living, to be used at the start of the experiment. Hence, in addition to the fact that this makes responses in small species easier to test, the method is also logistically simple and should therefore encourage the inclusion of more replicates and controls than what is current practice. Finally, as the Propagule Method does not involve any pretreatments specific for certain kinds of shell material (e.g., carbonate as needed in the calcein method; Bernhard et al. 2004) or require transparent shells in order to see the living cell while isolating populations to be used in an experiment, it is relatively simple and cheap and has the potential to work for a diversity of benthic foraminifera irrespective of shell composition. For discussion of dispersal potential via propagules in foraminifera with different life styles and sizes, e.g., large-sized and attached forms, see Alve (1999) and Alve and Goldstein (2003).

The fact that the experimental assemblages are treated in their original (natural) sediments implies that the Propagule Method optimizes the ability to mimic their natural conditions and thereby maximize the possibility of isolating the effects of the target parameter(s). Together with the fact that these sediments contain juveniles (i.e., the propagule bank) of a wider range of species than the ones recorded at the sampling site using conventional processing procedures (e.g., the > 63 μm-fraction), it also implies that the method is well suited for testing effects of changing environmental conditions at the assemblage level (Goldstein and Alve 2011). The populations and characteristics of assemblages (e.g., abundances, diversity, species composition) grown under different conditions can then be compared. On the other hand, using original sediments introduces a potential limitation in the sense that the cultures are xenic, i.e., apart from the foraminiferal propagules, the sediment fine fraction contains an unknown microbiota of bacteria, very small metazoans and algae. The effects of these organisms are unknown; they may interact with the foraminifera as predators or prey, via competition for space or food, or by modifying the microenvironment. However, utilizing a reasonably homogenous fine sediment fraction and using replicates minimize this impact on results.

Our knowledge of benthic foraminiferal syn- and autecology is still limited and based largely on interpretations of field- and laboratory studies that do not necessarily provide us with the direct and most critical cause-effect relationships. Consequently, one problem at our present stage of knowledge is that because, for most species, we do not know the optimal growth conditions, it may be difficult to grow exactly the target species required for answering a specific scientific question. On the other hand, obtaining information about these growth conditions are among challenges that can be addressed using the Propagule Method.

1.4 Applications of the Propagule Method

It is well established that the distribution and abundance of benthic foraminifera in modern and ancient sediments provide information on environmental change over a range of temporal and spatial scales. However, the trustworthiness of our interpretations depends on our understanding of their biology and ecology; certain aspects of this understanding can best be addressed through experimental approaches. Although laboratory experiments can never completely mimic natural conditions, they give insight into possible ecological cause-effect relationships as well as the speed and nature of biotic processes. The Propagule Method can be applied to most ecological questions concerning benthic foraminifera but is particularly suitable for studies concerning dispersal, response to environmental pressure, biological trait analyses, and shell chemistry.

Knowledge about *dispersal* mechanisms and processes have implications for understanding and approaching fundamental questions in biogeography, evolution, and several geoscience disciplines. Indeed, Myers (1936) long ago mentioned transport of juveniles as a possible dispersal strategy and a recent review based on all available data on live/stained benthic foraminifera concluded that "Propagules are the most likely mechanism of dispersal of species along the margins and across an ocean as well as between oceans" (Murray 2013). An example supporting this is the occurrence of sessile foraminifera on petroleum platform underwater legs on the Louisiana shelf, Gulf of Mexico, probably originating from carbonate hardgrounds to the west (Sen Gupta and Smith 2013). So far, the Propagule Method has been particularly useful to prove the presence of propagules of "exotic" species beyond the distribution of their conspecific adults. Examples include the attached *Planorbulina mediterranensis* growing from sediment collected at 320 m water depth in the Skagerrak, North Sea (Alve and Goldstein 2010) and shelf species growing from Georgia, USA, intertidal mudflat-sediments (Goldstein and Alve 2011) showing that not all propagules are derived from the "resident" populations; some are allochthonous. The latter study also illustrated how the method can aid identifying that a morphospecies (here *Miliammina fusca*) can consist of a complex of cryptic species with different environmental adaptations. The Propagule Method also shows how dispersal limitation, temperature, and salinity function in structuring foraminiferal associations (Goldstein and Alve 2011) and provides insight into dispersal distance and potential source populations (Alve and Goldstein 2010).

In order to improve our ability to recognize and interpret paleoenvironmental change, whether naturally or human induced, we need to better understand how foraminifera *respond to environmental pressure*, particularly the effects of single environmental parameters. This should not only include responses by a single species, but also changes in community composition and structure. The Propagule Method is an appropriate approach to such studies in that it provides an experimental design in which one parameter can be varied while holding others constant. For example, Goldstein and Alve (2011) were able to assess the effects of several parameters (temperature, salinity, sediment source) independently on the process of community assembly.

Understanding the significance of *biological traits* in benthic foraminifera (e.g., life strategy and duration, feeding, microhabitat, morphology) and how the traits are linked to environmental gradients is another aspect influencing their usefulness in both present and past environmental studies. So far, propagule experiments have demonstrated, e.g., opportunistic behavior in *Textularia earlandi* and *Haynesina germanica* (Alve and Goldstein 2010; Goldstein and Alve 2011). Through more frequent sampling during experiments, the method also has the potential to record changes in test size over time and thereby provide insight into species' growth and reproduction rates.

Quantitative paleoceanographic reconstructions rely heavily on knowledge of what controls the isotopic and elemental composition of calcareous benthic foraminiferal *shell chemistry*. While empirical calibrations based on field studies are important, they are also impacted by changes in environmental and ecological factors. Consequently, complementary geochemical information from populations grown in controlled laboratory experiments is needed to isolate effects of particular parameters (e.g., Filipsson et al. 2010). The Propagule Method is well suited for this kind of experiment because all shell material, except the thin-shelled proloculus (or first few chambers) of the first grown generation, will calcify under predetermined experimental conditions.

1.5 Summary and Conclusions

Benthic foraminiferal propagules (small juveniles) are abundant and diverse, they form "propagule banks" in marine sediments beyond the distribution of their conspecific adults, and start growing when exposed to appropriate conditions. These characteristics are applied experimentally in the "Propagule Method". Using the fine grained fraction of natural sediment, populations are grown (the size selected depends on the objective of the study) in microcosms under controlled conditions. Consequently, the Propagule Method optimizes the ability to mimic the species' natural conditions and thereby maximize the possibility of isolating the effects of the target parameter(s). The fine fraction of original (natural) sediments is isolated in static (closed) microcosms and exposed to experimental conditions. Growth commences in species for which these conditions are appropriate. At the conclusion of the experiment the contents of the microcosms are sieved using a screen with larger openings than that used initially to isolate the fine sediment. This removes the surrounding sediment and ensures that foraminifera harvested have grown and therefore responded positively to the experimental conditions. Advantages using the Propagule Method include that it

- ensures that all populations (live or dead) harvested from the microcosms at the end of experiments have responded positively to the treatment,
- focuses on the critical, juvenile developmental stages,
- is logistically simple,

- is suited to test traits and responses in small as well as larger species,
- has the potential to work for a diverse array of benthic foraminifera irrespective of shell composition,
- is well suited for testing effects of changing environmental conditions at the assemblage level under controlled conditions (i.e., different assemblages grown from the same propagule bank),
- allows investigation of how dispersal limitation plays a role in structuring foraminiferal associations, and
- can identify distributions of "exotic" or allochthonous species (i.e., species that grow in experimental assemblages but are absent from growth-stages of naturally occurring assemblages at that site).

Growing assemblages of foraminifera from their propagule banks under controlled conditions improves our understanding of the ecological requirements of individual taxa and how foraminiferal communities may respond to changing environments.

Acknowledgements We thank John W. Murray and Christopher J. Duffield for useful comments on the manuscript. This work was supported by National Science Foundation grant OCE 0850505 to STG.

References

Alve E (1999) Colonization of new habitats by benthic foraminifera: a review. Earth Sci Rev 46:167–185

Alve E, Bernhard JM (1995) Vertical migratory response of benthic foraminifera to controlled oxygen concentrations in an experimental mesocosm. Mar Ecol Prog Ser 116:137–151

Alve E, Goldstein ST (2002) Resting stage in benthic foraminiferal propagules: a key feature for dispersal? Evidence from two shallow-water species. J Micropalaeontol 21:95–96

Alve E, Goldstein ST (2003) Propagule transport as a key method of dispersal in benthic foraminifera. Limnol Oceanogr 48:2163–2170

Alve E, Goldstein ST (2010) Dispersal, survival and delayed growth of benthic foraminiferal propagules. J Sea Res 63:36–51

Bernhard JM (2000) Distinguishing live from dead foraminifera: methods review and proper applications. Micropaleontology 46:38–46

Bernhard JM, Blanks JK, Hintz CJ, Chandler GT (2004) Use of the fluorescent marker calcein to label foraminiferal tests. J Foraminiferal Res 34:96–101

Chandler GT, Williams DF, Spero HJ, Ziaodong G (1996) Sediment microhabitat effects on carbon stable isotopic signatures of microcosm-cultured benthic foraminifera. Limnol Oceanogr 41:680–688

de Nooijer LJ, Reichart GJ, Dueñas-Bohórquez A, Wolthers M, Ernst SR, Mason PRD, van der Zwaan B (2007) Copper incorporatioin in foraminiferal calcite: results from culturing experiments. Biogeosciences 4:493–504

Dissard D, Nehrke G, Reichart G-J, Bijma J (2010) The impact of salinity on the Mg/Ca and Sr/Ca ratio in the benthic foraminifera *Ammonia tepida*: results from culture experiments. Geochim Cosmochim Acta 74:928–940

Duguay LE (1983) Comparative laboratory and field studies on calcification and carbon fixation in foraminiferal–algal associations. J Foraminiferal Res 13:252–261

Ernst S, van der Zwaan B (2004) Effects of experimentally induced raised levels of organic flux and oxygen depletion on a continental slope benthic foraminiferal community. Deep-Sea Res 51:1709–1739

Ernst S, Morvan J, Geslin E, Le Bihan A, Jorissen FJ (2006) Benthic foraminiferal response to experimentally induced *Erika* oil pollution. Mar Micropaleontol 61:76–93

Filipsson HL, Bernhard JM, Lincoln SA, McCorkle DC (2010) A culture-based calibration of benthic foraminiferal paleotemperature proxies: δ18O and Mg/Ca results. Biogeosciences 7:1335–1347

Goldstein ST (1997) Gametogenesis and the antiquity of reproductive pattern in the Foraminiferida. J Foraminiferal Res 27:319–328

Goldstein ST (1999) Foraminifera: a biological overview. In: Sen Gupta BK (ed) Modern foraminifera. Kluwer, Dordrecht, pp 37–55

Goldstein ST, Alve E (2011) Experimental assembly of foraminiferal communities from coastal propagule banks. Mar Ecol Prog Ser 437:1–11

Grell KG (1956) Der Kerndualismus der Foraminifere *Glabratella sulcata*. Zeitschrift für Naturforschung 11b:366–368

Gustafsson M, Dahllof I, Blanck H, Hall P, Molander S, Nordberg K (2000) Benthic foraminiferal tolerance to tri-n-butyltin (TBT) pollution in an experimental mesocosm. Mar Pollut Bull 40:1072–1075

Hallock P (1981) Light dependence in *Amphistegina*. J Foraminiferal Res 11:40–46

Hemleben C, Kitazato H (1995) Deep-sea foraminifera under long time observation in the laboratory. Deep-Sea Res I 42:827–832

Hintz CJ, Chandler GT, Bernhard JM, McCorkle DC, Havach SM, Blanks JK, Shaw TJ (2004) A physicochemically-constrained seawater culturing system for production of benthic foraminifera. Limnol Oceanogr Methods 2:160–170

Hintz CJ, Shaw TJ, Chandler GT, Bernhard JM, McCorkle DC, Blanks JK (2006a) Trace/minor element:calcium ratios in cultured benthic foraminifera. Part I: inter-species and inter-individual variability. Geochim Cosmochim Acta 70:1952–1963

Hintz CJ, Shaw TJ, Chandler GT, Bernhard JM, McCorkle DC, Blanks JK (2006b) Trace/minor element:calcium ratios in cultured benthic foraminifera. Part II: ontogenetic variation. Geochim Cosmochim Acta 70:1964–1976

Hofker J (1951) The foraminifera of the Siboga Expedition, Part III. In: Brill EJ (ed), Ordo Dentata, sub-Ordines Protoforaminata, Biforaminata, Deuteroforaminata, Leiden, 513 p

Höglund H (1947) Foraminifera in the Gullmar Fjord and the Skagerak. Zoologiska bidrag från Uppsala 26:1–328

Kitazato H (1988) Locomotion of some benthic foraminifera in and on sediments. J Foraminiferal Res 18:344–349

Lister JJ (1895) Contributions to the life-history of the foraminifera. Philos Trans R Soc Lond Ser B 186:401–453

Lopez E (1979) Algal chloroplasts in the protoplasm of three species of benthic foraminifera: taxonomic affinity, viability, and persistence. Mar Biol 53:201–211

Mojtahid M, Zubkov MV, Hartmann M, Gooday AJ (2011) Grazing of intertidal benthic foraminifera on bacteria: assessment using pulse-chase radiotracing. J Exp Mar Biol Ecol 399:25–34

Murray JW (1991) Ecology and paleoecology of benthic foraminifera. Longman Scientific and Technical & Wiley, Harlow, 397 p

Murray JW (2000) The enigma of the continued use of total assemblages in ecological studies of benthic foraminifera. J Foraminiferal Res 30:244–245; 2002, ibid., 32: 200; (erratum)

Murray JW (2006) Ecology and applications of benthic foraminifera. Cambridge University Press, Cambridge, 426 p

Murray JW (2013) Living benthic foraminifera: biogeographical distributions and the significance of rare morphospecies. J Micropalaeontol 32:1–58

Murray JW, Alve E (2000) Major aspects of foraminiferal variability (standing crop and biomass) on a monthly scale in an intertidal zone. J Foraminiferal Res 3:177–191

Myers EH (1935) The life history of *Patellina corrugata* Williamson, a foraminifer. Bull Scripps Inst Oceanogr Tech Ser 3:355–392

Myers EH (1936) The life cycle of *Spirillina vivipara* Ehrenberg, with notes on morphogenesis, systematics and the distribution of foraminifera. J R Microsc Soc 56:120–146

Nomaki H, Heinz P, Hemleben C, Kitazato H (2005) Behavior and response of deep-sea benthic foraminifera to freshly supplied organic matter: a laboratory feeding experiment in microcosm environments. J Foraminiferal Res 35:103–113

Olsgard F (1999) Effects of copper contamination on re-colonisation of subtidal marine soft sediments—an experimental field study. Mar Pollut Bull 38:448–462

Pawlowski J, Lee JJ (1992) The life-cycle of *Rotaliella elatiana* n.sp.: a tiny macroalgavorous foraminifer from the Gulf of Elat. J Protozool 39:131–143

Pineda MC, McQuaid CD, Turon X, López-Legentil S, Ordóñez V, Rius M (2012) Tough adults, frail babies: an analysis of stress sensitivity across early life-history stages of widely introduced marine invertebrates. PLoS ONE 7:e46672. doi:10.1371/journal.pone.0046672

Röttger R (1976) Ecological observations on *Heterostegina depressa* (Foraminifera, Nummulitidae) in the laboratory and in its natural habitat. Mar Sed Spec Publ 1:75–80

Sen Gupta BK, Smith LE (2013) Foraminifera of petroleum platforms, Louisiana shelf, Gulf of Mexico. Mar Micropaleontol. doi:10.1016/j.marmicro.2013.01.001

Topping JN, Murray JW, Pond DW (2006) Sewage effects on the food sources and diet of benthic foraminifera living in oxic sediment: a microcosm experiment. J Exp Mar Biol Ecol 329:239–250

Toyofuku T, Kitazato H (2005) Micromapping of Mg/Ca values in cultured specimens of the high-magnesium benthic foraminifera. Geochem Geophys Geosyst 6 (Q11P05). doi:10.1029/2005GC00961

Wilson-Finelli A, Chandler GT, Spero HJ (1998) Stable isotope behavior in paleoceanographically important benthic foraminifera: results from microcosm culture experiments. J Foraminiferal Res 28:312–320

Young CM (1995) Behavior and locomotion during the dispersal phase of larval life. In: McEdward L (ed) Ecology of marine invertebrate larvae. CRC, Boca Raton, pp 249–278

Chapter 2
The Natural Laboratory of Algal Symbiont-Bearing Benthic Foraminifera: Studying Individual Growth and Population Dynamics in the Sublittoral

Johann Hohenegger, Antonino Briguglio, and Wolfgang Eder

Abstract Reproduction period, longevity and the chamber-building rate of symbiont-bearing benthic foraminifera, which are important for population dynamic studies can be estimated from field data. Laboratory investigations changing the grade of ecological variables cannot substitute for the complexity of natural conditions. Therefore, methods are developed, especially for the deeper sublittoral species, to estimate reproduction, lifespan and the individual growth rate under natural conditions for demonstrating the influence of environmental parameters.

Keywords Longevity • Natural growth • Reproduction • Standardization • Sublittoral sampling

2.1 Introduction

Investigations on the biology of algal symbiont-bearing benthic foraminifera (in the following shortened as SBBF) living in the eulittoral and upper sublittoral were predominantly based on laboratory studies. Soon after starting in the early fifties of the twentieth century, laboratory experiments on SBBF culminated in the works of Rudolf Röttger and co-workers, where reproduction and growth were studied in *Heterostegina depresssa*, the flagship of laboratory investigations on SBBF (e.g. Röttger 1972, 1976; Röttger and Spindler 1976; Röttger et al 1980),

J. Hohenegger (✉) • W. Eder
Department of Palaeontology, University of Vienna, 1090 Vienna, Austria
e-mail: johann.hohenegger@univie.ac.at

A. Briguglio
Department of Palaeontology, University of Vienna, 1090 Vienna, Austria

Natural History Museum Vienna, Burgring 7, 1010 Vienna, Austria

H. Kitazato and J.M. Bernhard (eds.), *Approaches to Study Living Foraminifera: Collection, Maintenance and Experimentation*, Environmental Science and Engineering, DOI 10.1007/978-4-431-54388-6_2, © Springer Japan 2014

followed by *Cycloclypeus carpenteri* (Krüger 1994; Lietz 1996), *Nummulites venosus* (Krüger 1994), *Calcarina gaudichaudii* (Röttger et al. 1990) and *Amphistegina lessonii* (Dettmering 1997). Further important studies on laboratory cultures of SBBF were performed by John J Lee (e.g. Lee et al. 1980; Lee et al. 1991), Pamela Hallock and co-workers (e.g. Toler and Hallock 1998; Toler et al. 2001), while Kazuhiko Fujita and Sven Uthike, both with co-workers, recently conducted laboratory experiments under different environmental conditions (e.g. Fujita and Fujimura 2008; Hosono et al. 2012; Uthicke and Fabricius 2012; Uthicke et al. 2012).

Field observations on the biology of SBBF concentrate on eulittoral species living on the reef crest (Sakai and Nishihira 1981; Hohenegger 2006) or in regions of the shallowest sublittoral (e.g. Zohary et al. 1980; Fujita and Hallock 1999; Fujita et al. 2000; Fujita 2004; Osawa et al. 2010; Uthicke and Altenrath 2010; Reymond et al. 2011; Ziegler and Uthicke 2011). Observation on SBBF in the deeper sublittoral is more difficult due to intense hydrodynamics that hinder secure fixing of technical equipment for studies in the natural environment.

For the investigation of reproductive timing, growth and longevity of generations (agamonts, gamonts and schizonts) in SBBF, the chamber-building rate is of primary importance. The chamber-building rate represents the independent character for measuring the influence of time-dependent environmental factors like spring tides or seasonality in reproduction and growth. These environmental factors cause changes in temperature, solar irradiation, water transparency, input of inorganic and organic nutrition etc.

Growth experiments in laboratory cultures cannot represent natural conditions, although the objective is to simulate them; thus they cannot give reliable information about reproduction time, growth, longevity and life cycles. Two examples may show these difficulties. In September 1992 *Peneroplis antillarum* was sampled from intertidal pools of the reef crest NW of Sesoko Island (Hohenegger 1994) and put into Petri-dishes. Only water was changed weekly using sea water from the upper sublittoral in front of the Sesoko Marine Laboratory. Except water movement, other factors influencing growth were kept as natural as possible because the Petri-dishes were exposed to natural sun light. After three months, the laboratory sample was compared with a sample taken from the same pool where the lab sample originated. The differences were striking; while the sample from the pool showed individuals with undisturbed growth that can be modelled by a logarithmic spiral, individuals kept in the laboratory showed restricted growth and several growth disturbances leading to deviations from the logarithmic spiral (Fig. 2.1).

Similar results have been observed comparing *Heterostegina depressa* tests from individuals collected from the natural environment with individuals kept under laboratory conditions. A comprehensive way to study and illustrate test morphology en toto in space is the use of computed tomography. More details on this technique, its uses and applications in foraminiferal biology and palaeontology are reported in two other papers contained in this book.

Two specimens have been scanned with a micro computer tomograph (microCT) at the Department of Palaeontology, University of Vienna. The very high scanning resolution (<4 μm) allowed the visualization and the quantification of almost any morphological parameter. In Fig. 2.2, a specimen (specimen's name: A1) of

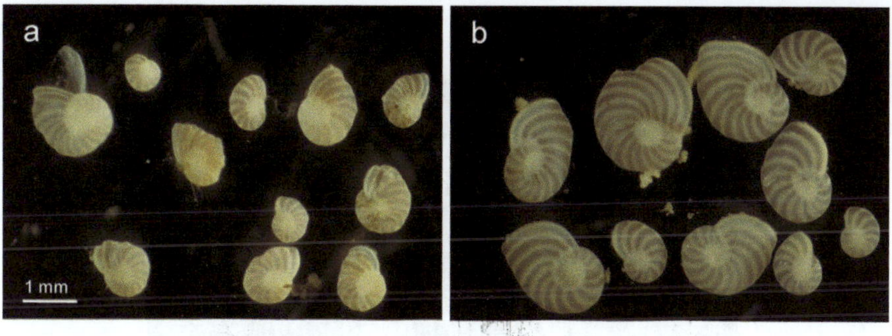

Fig. 2.1 Light microscope micrographs of *Peneroplis antillarum* (**a**) sampled in September 28, 1992 and kept living in the laboratory until December 5, 1992; (**b**) sampled in December 5, 1992 from the same sampling location

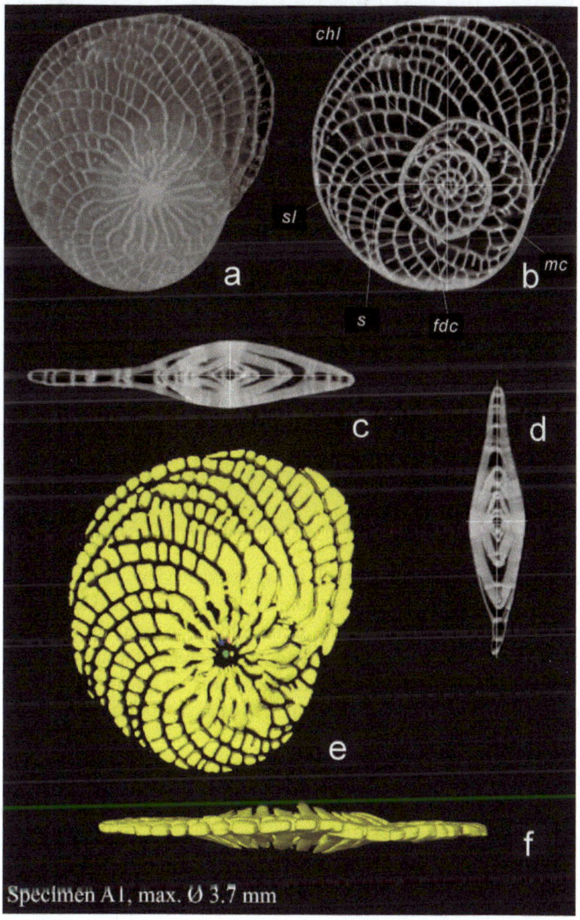

Fig. 2.2 MicroCT scans of *Heterostegina depressa* specimen A1, (**a**) external view of the 3D model, (**b**) equatorial section of the specimen, (**c**, **d**) axial section of the specimen along the axes visible in **b**, (**e**) equatorial view of all segmented chambers, operculinid chambers are visibible in the central part and are not yellow colored, (**f**) axial view of the segmented chambers. *fdc* first divided chamber, *mc* marginal chord (see marginal canal within), *s* setpum, *sl* septulum, *chl* chamberlet

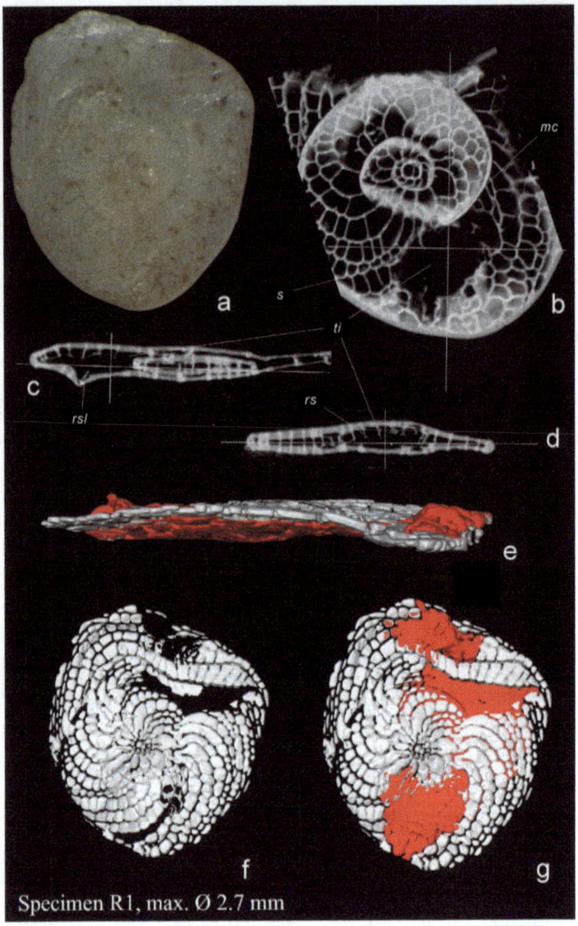

Fig. 2.3 *Heterostegina depressa* specimen R1, (**a**) external view of the specimen under microscope, (**b**) equatorial section of the specimen (note the large hole created at chamber 45), (**c, d**) axial section of the specimen along the axes visible in **b** (note the incomplete septula), (**e**) axial view of the segmented chambers (in *red*, the large hole which is extending through the test), (**f, g**) equatorial view of the segmented chambers (in **f** without the large hole, in **g** with the view of the large hole in *red*), *mc* marginal chord, *ti* test inflation, *s* septum, *rs* reduced septum, *rsl* reduced septulum

Heterostegina depressa was collected alive at 20 m depth in front of Sesoko Island (Hohenegger et al. 1999) and immediately dried. In Fig. 2.3, a specimen of the same species (specimen's name: R1) is displayed; it was cultivated under laboratory condition at the University of Kiel, Germany (Röttger 1972; Röttger and Spindler 1976; Krüger 1994). For each specimen, microCT slices, 3D model reconstructions and specimen segmentation are reported.

The shell morphology of specimen A1 represents the general and common shape of specimens belonging to this species: a slightly evolute spiral coiled test (Fig. 2.2a) with several undivided initial chambers, then followed by chambers divided into chamberlets formed by complete septula (Fig. 2.2b–d) and a centrally thickened shell becoming much flatter at the periphery by increasing the distance between the septa (Fig. 2.2e, f).

On the contrary, specimen R1 (Fig. 2.3), which seems to have a normal shape by observing it under the microscope, as surely Röttger did in his lab, shows several strong shell variations: the inconstant curvature of the septa (Fig. 2.3b), the sudden inflation of the test (Fig. 2.3b–d) and (connected to this morphology) the vertical reduction of septa and septula (Fig. 2.3e, f).

The thick marginal chord and the septa in Fig. 2.2b, where both marginal and septal canal systems are visible, are completely lost in specimen R1 (Fig. 2.3b), where the septa appear to be made of compounded bulging singular septula. This pattern starts to emerge after approximately the 10th chamber in R1. Furthermore, on the axial slices of specimen R1 (Fig. 2.3c, d), taken along the axes visible on the equatorial section (Fig. 2.3b), a big cavity or an inflation of the test walls instead of a normal chambered internal structure is visible (Fig. 2.3e–g, in red). This hollow space starts around the 45–47th chamber, but early chambers also show a trend of septal reduction resulting in complex chamberlet geometries. Coincidental to the forming of this cavity, specimen R1 shows a vertical reduction of the septula, which extend neither to the following septum nor laterally to the test wall (Fig. 2.3c, d).

Possible explanations for such abnormal growth must be connected to particular culturing conditions, which according to the published material was very advanced for the late 1970s, but still not representative of natural conditions.

Therefore, it is necessary to study individual and population growth under natural conditions. Because the installation of technical equipment is difficult in regions with extreme hydrodynamics like the upper sublittoral and often equipment can be destroyed by tropical storms, investigation by sampling in more or less constant intervals over a time period of at least one year is a possible solution. In the following, sampling, data collection and evaluation will be demonstrated.

2.2 Sampling

The best sampling method in the upper sublittoral from 5 to 60 m depth is by SCUBA diving. The sampling procedure is described in detail below.

a. The determination of *location*, *water depth* and *sedimentary conditions* must be based on former investigations about the regional distribution and abundance maximum of the species if interest. For example, investigations by Hohenegger (2004) NW of Sesoko Island, Okinawa, Japan, demonstrated optimum conditions for *Palaeonummulites venosus* at 50 m depth on sandy bottom in the northern transect, while *Heterostegina depressa* has its optimum at 20 m depth on firm substrates, which are structured coral rock and boulders.
b. A *sampling interval* must be selected, but actual sampling events will typically depend on weather conditions, which may hinder consistent intervals. Irregular sampling intervals may range from weeks to months, where the latter represents

the upper limit, because larger intervals could obliterate the data set. To investigate the influence of tides, weekly sampling close to spring and neap tides is necessary.

c. To make measures on environmental conditions—especially irradiation—comparable over seasons, noon should be the *time of day* for *sampling*.
d. *Sampling methods* depend on the substrate. For *soft substrates* with grain sizes of pebble to clay according to the Udden-Wenworth grain size classification (Boggs 2006), a prismatic plastic box with a secure lid should be used. At the sampling point, the sediment has to be dredged into the box, where only the upper 2 to 3 cm of the sediment should be taken as the maximum dredging depth. Afterwards, the box must be closed by securing the lid.

   For investigating species living on *firm substrates*, boulders and cobbles must be gathered and put into closable plastic carrier bags.

   Additional *water samples* should be taken for investigating the chemical composition of seawater in the laboratory.
e. The *number of sampling points* at the location must be ≥4, randomly distributed and in considerable distance from each other (approximately 5 m) for smoothing the effects of patchy distributions.
f. During sampling, on-site measurements of physical factors like *temperature* and *light intensity* (= irradiance) should be measured at the sampling location. Irradiance must also be measured at the surface, because *relative irradiance*, which is independent of weather conditions, relates irradiance at the sampling depth to sea surface irradiance by

$$irrad_{rel} = \frac{\ln\left(irrad_{sample}\right)}{\ln\left(irrad_{sea\ surface}\right)} \tag{2.1}$$

   where the unit of irradiance corresponds to

$$6 \cdot 10^{17}\, photons\, \mathrm{m}^{-2}\, \mathrm{sec}^{-1} = 1\, microEinstein\left(\mu E\right)\mathrm{m}^{-2}\, \mathrm{sec}^{-1}$$

   Logarithms must be used, because irradiation follows an exponential decrease (Hohenegger et al. 1999). Using relative irradiation, changes in light intensities over seasons due to inorganic or organic input can be calculated independent of weather conditions on the sampling date.

For studying individual and population growth in species with their distribution optimum deeper than 60 m, sampling by SCUBA diving is difficult or impossible. Sampling in the deeper sublittoral can be performed by *crab-sampling*, *coring* or *dredging*. While individual growth over the year can be measured using all sampling methods, population growth that needs a standardized substrate surface can be estimated using crab sampling or coring, where the bottom surface area is either determined by the core diameter or can be approximated by measuring the opening of the crab sampler. Area determination of a dredged bottom surface is difficult to impossible using dredgers with unfixed penetration depth.

2.3 Concentrating Living Individuals

Picking living foraminifera out of the samples depends on substrate conditions. This should be done in the laboratory using vessels filled with sea water. Vessel size should not be too deep making investigation with a binocular microscope impossible.

a. *Boulders* and *cobbles* must be intensively cleaned over the investigation vessel using a dental brush. Afterwards, specimens that could not be removed by brushing must be picked under the binocular and put into the investigation vessel. Cleaned boulders and cobbles should be washed with freshwater, dried and stored for further investigations, or returned to seawater and returned to the collection site, if required by local regulations.

   Soft sediment is directly placed into the investigation vessel filled with sea water. Because the thickness of the sediment layer in the vessel should not exceed 2 mm, only parts of the sampled sediment can be put into the vessel. The proportion depends on vessel area that must be equally covered by the sediment. To extract fine organic and fluffy material that could be abundant in fine-grained sediment, repeat decantation using sea water is necessary. Decantation of fine silt to mud from sandy sediments is also necessary. This fine fraction should be put in separate vessels for grain size analysis.

   Depending on vessel area and sample size, soft sediments require several investigation vessels simultaneously.
b. Investigation vessels should *rest for one day* at least, because living foraminifera, retract their colored protoplasm due to disturbance by sampling and preparation, but will refill the final chambers during this calm resting period.
c. Living foraminifera can now be *picked* out using fine and flexible forceps and put into separate cups filled with sea water. The *identification of living* SBBF is easy compared to non symbiont-bearing foraminifera, because SBBF are colored by their symbiotic microalgae. Living peneroplids can be identified by their purple color (*Porphyridium* belonging to rhodophyta), archaiasinids and *Parasorites* by green colour (*Chlamydomonas* belonging to chlorophyta), soritids by olive to ochre colors (*Symbiodinium* belonging to dynophyta), while alveolinids and all hyaline SBBF are characterized by light ochre color caused through symbiotic bacillariophyta (diverse genera and species of diatoms).

   Difficulties for identifying living individuals may arise in hyaline SBBF, especially nummulitids, because they can be colored by bacteria or other non-symbiotic microalgae after leaving the test by reproduction or death. In contrast to living individuals, which are evenly colored by light ochre, coloring of empty tests is more intense, spotty or dark-stained.
d. After picking all living individuals, the remaining sediments must be washed in freshwater, dried and stored for further investigations.
e. The picked living individuals are now ready for further investigations, either becoming the base for laboratory experiments or for investigating individual and population growth. For studying the chamber building rate, individuals of the species of interest should be washed in freshwater, dried and stored.

2.4 Determination of Environmental Parameters

On-site measurements at the sampling station over the sampling period provide information about changes in *temperature* and relative *irradiance*.

The crossing of *storms* must be recorded, because they can intensively disturb the bottom surface down to 100 m water depth, especially entraining and transporting foraminifera living on soft substrate. Disturbances by storms can be expressed either in the composition of the foraminiferal fauna or in distinct changes of grain size. Disturbance in the faunal composition was noted in a sample collected in 1992 from 50-m water depth that was taken after the first seasonal crossing of a typhoon. In contrast to samples from 50 m taken before the typhoon season, abundant living *Peneroplis pertusus*, *P. antillarum*, *Dendritina ambigua* and *D. zhengae*, which are typically restricted to the shallowest sublittoral (<30 m), were found at this depth, obviously having been transported to deeper sites. Therefore, *grain size analysis* is necessary for each sampling location and event.

Water taken from the sampling station should be investigated in the laboratory just after sampling. Chemical parameters characterizing the environment of the sampling station like pCO_2, pH, nitrate concentration, O_2 and the organic carbon content should be measured.

2.5 Investigating Individual Growth and Population Dynamics

2.5.1 *Measurements*

For SBBF, only three simple measurements are necessary for the determination of chamber-building rate and population dynamics. These are the number of individuals n, number of chambers m_i and the largest diameter d_i of individual i.

The determination of a standardized in situ surface area of the sample, which is necessary for population dynamic investigations, depends on the sampling method. Using cores, sample surface a^2 is given by the inner core diameter d_{core}

$$a^2_{observed} = \left(d_{core} / 2\right)^2 \pi. \tag{2.2}$$

This area can be approximated in crab sampling by the opening size of the sampler. Cores can be taken by SCUBA.

The area of firm substrates like cobbles and rubble exposed to the water column can be measured using *image analysis*.

For soft sediments, the volume v of the dried sediment can be obtained using a *graduated measurement vessel*. Presuming a mean dredging depth l of the box used by the diver, sampling surface a^2 can be calculated by

$$a^2_{observed} = v / l \tag{2.3}$$

Now, the standardized individual number n^* for sample j is given by

$$n_j^* = \frac{a_{theoretical}^2}{a_{observed}^2} n_j \tag{2.4}$$

Standardization is always necessary for comparing samples of different sizes.

2.5.2 Statistical Investigation

Chamber number m_{ij} and size d_{ij} of n_j individuals of the sample j can be used to determine reproduction period, longevity and chamber-building rate of the species of interest.

Peneroplis antillarum sampled from tidal pools on the reef crest NW of Sesoko-Island between September 1992 and August 1993 will be used as an example (Hohenegger 2006). Sampling was performed in different intervals depending on weather condition and spring tides trying to approximate monthly sampling. Because the first samples from September and October 1992 were not treated in the requested manner, data processing started with the beginning of December 1992 (Fig. 2.4).

To demonstrate how reproduction, growth and longevity can be estimated, the largest diameter of *P. antillarum* shells was used.

a. First, a *frequency graph* must be constructed for every sample. In case of size measurements, the *histogram* is particularly useful for graphical representation of the frequency distribution. The lower and upper limit of the measurement scale must be identical for all histograms and interval width must be the same for all sample histograms, whereby the number of intervals should not exceed the square root of the largest sample. For comparison of frequencies, the abundance scale must be equal for all samples (Fig. 2.4).

   Contrary to size, which is a continuous variable, a *bar diagram* is the correct graph to show frequencies of the meristic (= natural numbers including 0) character chamber number (Fig. 2.5),
b. When in situ sampling surface areas are different, class abundance in the histogram or in the bar diagram must be *standardized* according to Eq. (2.4).
c. Using size in species with shells that can be modelled by a logarithmic spiral (like *Operculina*, *Planoperculina*, *Planostegina*, *Palaeonummulites* and *Heterostegina* in sublittoral SBBF), the histograms are always left-side skewed (like in the eulittoral *P. antillarum*; Fig. 2.4). To get normal-distributed histograms the transformation of original measurements into logarithms is necessary

$$d_{ij}^* = \ln d_{ij} \tag{2.5}$$

d. The *upper limit* of *minimum* size and/or chamber number characterizing the *offspring* has to be determined by investigating the size of offspring obtained in laboratory investigations or measuring the size of the embryonal apparatus (in *P. antillarum* all individuals smaller than 0.3 mm). Thereafter, *offspring frequencies* can be obtained for all samples using the cumulative frequency distributions up to this limit (Fig. 2.4).

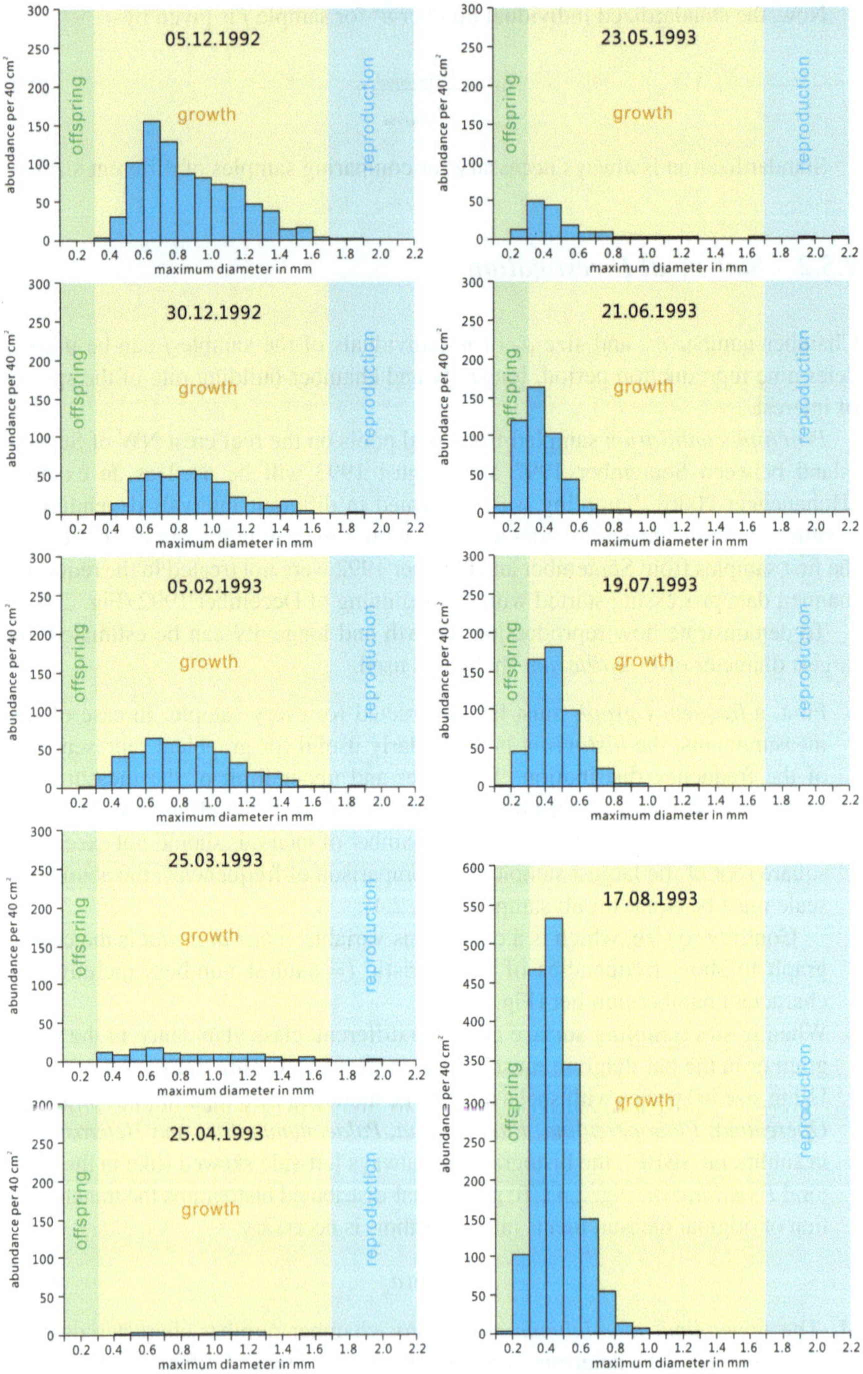

Fig. 2.4 *Peneroplis antillarum*: Histograms of test size measured as the largest diameter; test size <0.3 mm characterizing offspring and test size >1.6 mm characterizing individuals ready for reproduction are marked

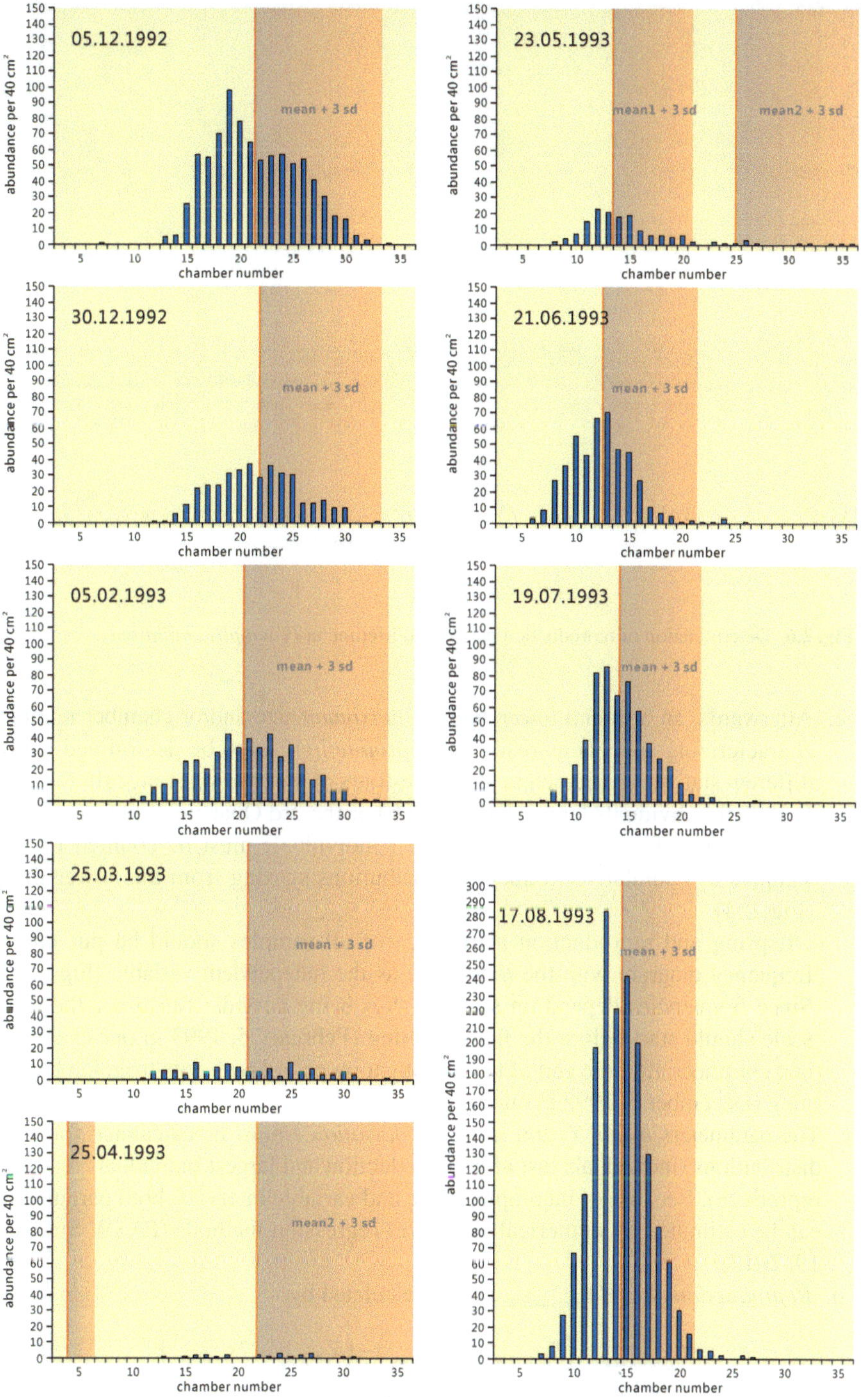

Fig. 2.5 *Peneroplis antillarum*: Bar diagrams of chamber numbers; the maximum chamber number defined as the arithmetic mean plus 3 times the standard deviation (Eq. (2.10))

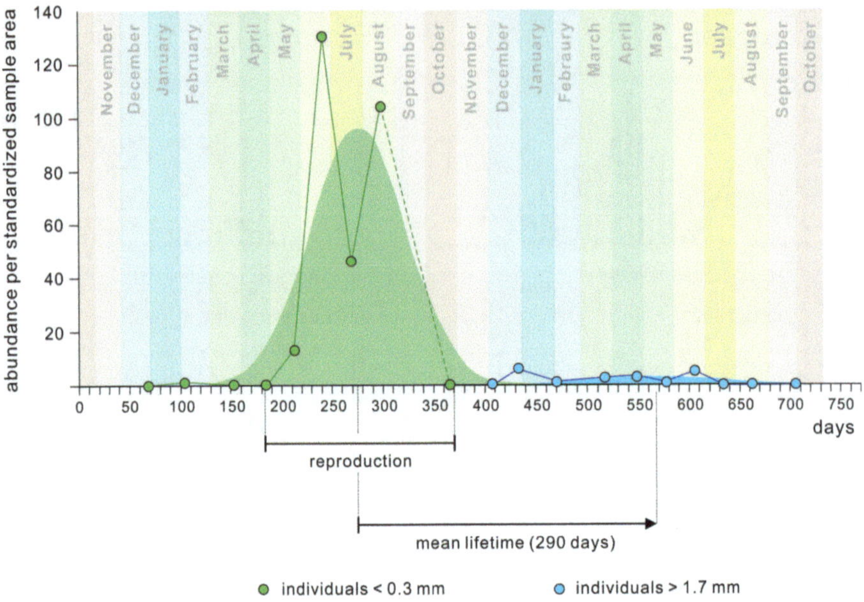

Fig. 2.6 Determination of reproduction period and lifetime in *Peneroplis antillarum*

e. Afterwards, an artificial *lower limit* of *maximum* size and/or chamber number characterizing specimens ready for *reproduction* has to be determined using different statistical parameters like three-fours of the total size range (in *P. antillarum* all individuals larger than 2.6 mm) or the 3rd Quartile of the sum of all distributions. The abundance of largest individuals must be counted for all samples as cumulative frequency distributions starting from this lower limit (Fig. 2.4).
f. Offspring and reproduction abundance of all samples should be put into a frequency diagram with the *time scale* as the independent variable (Fig. 2.6). Since frequencies depend on seasons, thus being periodic functions, the time scale should start before the first offspring (February 5, 1993 in our example) and continues after the end of the investigation period with data from the beginning (December 5, 1992 in our example; Fig. 2.4).
g. The parameters *mean* $\overline{x}$ and *standard deviation* s must be calculated for both distributions (individuals just after reproduction and largest individuals ready to reproduce). Because of incomplete data and variable intervals, both parameters can be estimated by numerical (iterative) regression methods (PASW Statistic 19, 2010).
h. *Reproduction period* $t_{reproduction}$ can be calculated by

$$t_{reproduction} = \left(\overline{x}_{offspring} + 2s_{offspring}\right) - \left(\overline{x}_{offspring} - 2s_{offspring}\right) \tag{2.6}$$

because 96 % of observations are positioned within this interval (Fig. 2.6).

In our example, the reproduction period is from the end of April until the end of October, peaking in July (Fig. 2.6).

Table 2.1 Statistical parameters of chamber numbers for calculating the chamber building rate

Datum	Days	Mean chamber number	Standard deviation	Coefficient of variation	Maximum chamber number
4/25/1993	1	4.0	0.81		6.4
5/23/1993	28	13.5	2.49	5.41	21.0
6/21/1993	57	12.4	3.01	4.13	21.5
7/19/1993	85	14.2	2.89	4.92	22.8
8/17/1993	114	14.4	2.70	5.32	22.5
10/24/1992	182	22.3	3.36	6.63	32.3
12/5/1992	224	21.5	4.13	5.21	33.9
12/30/1992	249	21.6	3.97	5.44	33.5
2/5/1993	286	20.9	4.42	4.71	34.1
3/25/1993	334	20.8	5.46	3.81	37.2
4/25/1993	366	22.1	4.93	4.48	36.9
5/23/1993	394	24.8	5.46	4.55	41.2

i. An averaged *lifetime* t_{life} can be calculated by

$$t_{life} = \overline{x}_{reproduction} - \overline{x}_{offspring}. \qquad (2.7)$$

In our example the averaged lifetime is 290 days (Fig. 2.6). Measuring lifetime using the interval between the first offspring (February 5) and the first individuals indicating reproduction (December 5) last for 302 days, similar to the estimation by the distance of means. Therefore, the lifetime of *P. antillarum* can be estimated as approximately one year.

Calculation of the *chamber building rate* is more complex depending on the reproduction period and longevity. The following steps are proposed:

j. Find sample j, where the first offspring during the investigation period appears ($j=1$; May 23 in our example, Fig. 2.4). Start the investigation period $t=1$ (in days) with the datum of the sample just before the first offspring sample (April 25 in our example, Fig. 2.4).

k. Calculate the *mean* $\overline{x}_j$ and *standard deviation* s_j of chamber number for all samples j (Table 2.1).

l. Calculate the **coefficient of variation**

$$CV_j = \overline{x}_j / s_j \qquad (2.8)$$

for all samples and check their constancy by linear regression analysis. In case of significant constancy calculate the averaged coefficient of variation, otherwise calculate the regression coefficients (Table 2.1).

m. Set the time of initial sampling period ($j=0$; April 25 in our example), which corresponds to the sample just before $j=1$, as $t_0=1$. The chamber number m_0 is based on the chamber number $m_{offspring}$ of the offspring grown in the laboratory. In *Peneroplis antillarum*, this chamber number is 3. Since most foraminifera build the following chamber within one day (Röttger 1972; Krüger 1994; Lietz 1996), the chamber number of the first day after offspring becomes

$$m_0 = m_{offspring} + 1 \qquad (2.9)$$

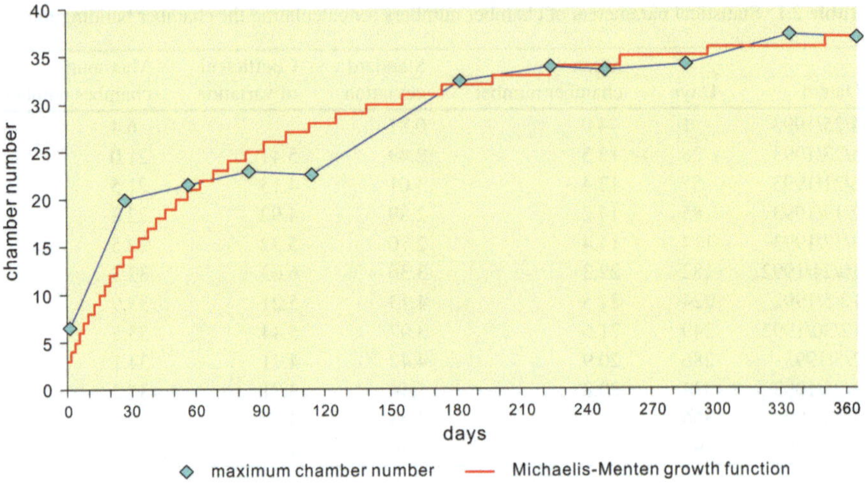

Fig. 2.7 Determination of the chamber building rate using the Michaelis–Menten growth function based on the maximum chamber numbers (Fig. 5)

which is 4 in *Peneroplis antillarum.*

n. Calculate the upper distribution limit for all samples by

$$m_{j\max} = \overline{x}_j + 3s_j \tag{2.10}$$

To get the upper limit for the initial sample $j=0$, the necessary standard deviation can be calculated by

$$s_0 = m_0 / CV_{mean} \tag{2.11}$$

o. Based on the relation between time t in days and the maximum chamber number $m_{\max}$, the chamber building rate can be calculated using the Michaelis–Menten function

$$t = am_{\max} / \left(b + m_{\max}\right) \tag{2.12}$$

with the reverse function

$$m_{\max} = bt / \left(a - t\right) \tag{2.13}$$

To standardize this function by the chamber number of the offspring at $t = 0$, the value of function 2.12 at $m_{offspring}$ must be subtracted from the function values t of Eq. (2.12):

$$t_j^* = am_{j\max} / \left(b + m_{j\max}\right) - am_{offspring} / \left(b + m_{offspring}\right) \tag{2.14}$$

The subtrahend in *Peneroplis antillarum* is 4.9, leading to the function

$$t = -67.619m / \left(-44.073 + m\right) - 4.9$$

p. The results of chamber building rate can now be represented as a function graph (Fig. 2.7).

These statistical methods can be used to determine chamber building rate, reproduction period and lifetime of algal SBBF in all environments by taking standardized samples in more or less regular intervals over at least one year.

References

Boggs S Jr (2006) Principles of sedimentology and stratigraphy, 4th edn. Pearson, Upper Saddle River

Dettmering C (1997) Untersuschungen zur Biologie von Großforaminiferen der Gattung *Amphistegina*, Dissertation Universität Kiel

Fujita K (2004) A field colonization experiment on small-scale distributions of algal symbiont-bearing larger foraminifera on reef rubble. J Foraminiferal Res 34:169–179

Fujita K, Fujimura H (2008) Organic and inorganic carbon production by algal symbiont-bearing foraminifera on northwest Pacific coral-reef flats. J Foraminiferal Res 38:117–126

Fujita K, Hallock P (1999) A comparison of phytal substrate preferences of *Archaias angulatus* and *Sorites orbiculus* in mixed macroalgal-seagrass beds in Florida Bay. J Foraminiferal Res 29:143–151

Fujita K, Nishi H, Saito T (2000) Population dynamics of *Marginopora kudakajimensis* Gudmundsson (Foraminifera : Soritidae) in the Ryukyu Islands, the subtropical northwest Pacific. Mar Micropaleontol 38:267–284

Hohenegger J (1994) Distribution of living larger Foraminifera NW of Sesoko-Jima, Okinawa Japan. PSZN I Mar Ecol 15:291–334

Hohenegger J (2004) Depth coenoclines and environmental considerations of Western Pacific larger foraminifera. J Foraminiferal Res 34:9–33

Hohenegger J (2006) The importance of symbiont-bearing benthic foraminifera for West Pacific carbonate beach environments. Mar Micropaleontol 61:4–39

Hohenegger J, Yordanova E, Nakano Y, Tatzreiter F (1999) Habitats of larger foraminifera on the upper reef slope of Sesoko Island, Okinawa, Japan. Mar Micropaleontol 36:109–168

Hosono T, Fujita K, Kayanne H (2012) Estimating photophysiological condition of endosymbiont bearing *Baculogypsina sphaerulata* based on the holobiont color represented in CIE L*a*b* color space. Mar Biol 159:2663–2673

Krüger R (1994) Untersuchungen zur Entwicklung rezenter Nummulitiden: *Heterostegina depressa, Nummulites venosus* und *Cycloclypeus carpenteri*. Dissertation Universität Kiel, Uni Press, Hochschulschriften

Lee JJ, McEnery ME, Garrison JR (1980) Experimental studies of larger foraminifera and their symbionts from the Gulf of Elat on the Red Sea. J Foraminiferal Res 10:31–47

Lee JJ, Sang K, ter Kuile B, Strauss E, Lee PL, Faber WW Jr (1991) Nutritional and related experiments on laboratory maintenance of three species of symbiont-bearing, large foraminifera. Mar Biol 109:417–425

Lietz R (1996) Untersuchungen zur Individualentwicklung der Großforaminifere *Cycloclypeus carpenteri* Carpenter (1856). Dissertation, Universität Kiel

Osawa Y, Fujita K, Umezawa Y, Kayanne H, Ide Y, Nagaoka T, Miyajima T, Yamano H (2010) Human impacts on large benthic foraminifers near a densely populated area of Majuro Atoll, Marshall Islands. Mar Pollut Bull 60:1279–1287

PASW Statistic 19 (2010) Version 19.0.0, IBM

Reymond CE, Uthicke S, Pandolfi JM (2011) Inhibited growth in the photosymbiont-bearing foraminifer *Marginopora vertebralis* from the nearshore Great Barrier Reef, Australia. Mar Ecol Prog Ser 435:97–117

Röttger R (1972) Analyse von Wachstumskurven von *Heterostegina depressa* (Foraminifera: Nummulitidae). Mar Biol 17:228–242

Röttger R (1976) Ecological observations of *Heterostegina depressa* (Foraminifera, Nummulitidae) in the laboratory and in its natural habitat. Marit Sediments Spec Publ 1:75–79

Röttger R, Spindler M (1976) Development of *Heterostegina depressa* individuals (Foraminifera: Nummulitidae) in laboratory cultures. Marit Sediments Spec Publ 1:81–87

Röttger R, Irwan A, Schmaljohann R (1980) Growth of the symbiont–bearing foraminifera *Amphistegina lessonii* d'Orbigny and *Heterostegina depressa* d'Orbigny (Protozoa). In: Schwemmler W, Schenk HEA (eds) Endocytobiology, endosymbiosis and cell biology 1. De Gruyter, Berlin, pp 125–132

Röttger R, Krüger R, De Rijk S (1990) Larger foraminifera: variation in outer morphology and *Prolocular* size in *Calcarina gaudichaudii*. J Foraminiferal Res 20:170–174

Sakai K, Nishihira M (1981) Population study of the benthic foraminifer *Baculogypsina sphaerulata* on the Okinawa reef flat and preliminary estimation of its annual production. In: Proceedings of the fourth international coral reef symposium 2, Manila, pp 763–766

Toler SK, Hallock P (1998) Shell malformation in stressed *Amphistegina* populations: relation to biomineralization and paleoenvironmental potential. Mar Micropaleontol 34:107–115

Toler SK, Hallock P, Schijf J (2001) Mg/Ca ratios in stressed foraminifera, *Amphistegina gibbosa*, from the Florida Keys. Mar Micropaleontol 43:199–206

Uthicke S, Altenrath C (2010) Water column nutrients control growth and C/N ratios of symbiont-bearing benthic foraminifera on the Great Barrier Reef, Australia. Limnol Oceanogr 55: 1681–1696

Uthicke S, Fabricius KE (2012) Productivity gains do not compensate for reduced calcification under near-future ocean acidification in the photosynthetic benthic foraminifer species *Marginopora vertebralis*. Glob Change Biol 18:2781–2791

Uthicke S, Vogel N, Doyle J, Schmidt C, Humphrey C (2012) Interactive effects of climate change and eutrophication on the dinoflagellate-bearing benthic foraminifer *Marginopora vertebralis*. Coral Reefs 31:401–414

Ziegler M, Uthicke S (2011) Photosynthetic plasticity of endosymbionts in larger benthic coral reef Foraminifera. J Exp Mar Biol Ecol 407:70–80

Zohary T, Reiss Z, Hottinger L (1980) Population dynamics of *Amphisorus hemprichii* (Foraminifera) in the Gulf of Elat (Aqaba), Red Sea. Eclogae Geol Helv 73:1071–1094

Chapter 3
Methods for Estimating Individual Growth of Foraminifera Based on Chamber Volumes

Johann Hohenegger and Antonino Briguglio

Abstract Based on chamber volumes, different methods for evaluating growth of individual foraminifera are shown using the generalized logistic growth function or the Gompertz function. Residuals to the theoretical functions were calculated to differentiate between instantaneous or oscillatory deviations. The chamber-building rate must be calculated allowing inferences of time-dependent influences by environmental factors.

Keywords Chamber-building rate • Cross correlation • Growth functions • Oscillation • Residuals

3.1 Introduction

Cellular growth in multi-chambered (polythalamous) foraminifera can be studied based on chamber volumes, because the expanded cell protoplasm can fill completely the chambers of the shell. Therefore, the sum of chamber volumes approximates the volume of the living cell.

Cell growth is marked in multi-chambered foraminifera in a sequence of chamber constructions. Chamber form follows a morphological program that is genetically fixed leading to regularly-shaped shells. According to this morphogenetic program, many foraminiferal shells can be modelled (Labaj et al. 2003).

J. Hohenegger (✉)
Department of Palaeontology, University of Vienna, 1090 Vienna, Austria
e-mail: johann.hohenegger@univie.ac.at

A. Briguglio
Department of Palaeontology, University of Vienna, 1090 Vienna, Austria

Natural History Museum Vienna, Burgring 7, 1010 Vienna, Austria

H. Kitazato and J.M. Bernhard (eds.), *Approaches to Study Living Foraminifera: Collection, Maintenance and Experimentation*, Environmental Science and Engineering,
DOI 10.1007/978-4-431-54388-6_3,

The chamber-building rate, e.g. the time difference between constructions of consecutive chambers, can be calculated by the first derivative of the growth function $j=f(t)$, where j indicates chamber numbers depending on time t. Environmental conditions that weakly change or oscillate through the lifetime of the foraminifer do not influence the form of the function, but strong differences between environmental conditions are expressed in parameters describing the rate of increase. An instantaneous change in environmental conditions during a given lifetime leads to an abrupt change in the rate of increase or stops further growth.

Because the chamber-building rate seems to be correlated with chamber-volume growth (Röttger 1972; Röttger and Spindler 1976), all periodic or instantaneous deviations from averaged environmental conditions during a foraminiferan's lifetime like seasonal effects and other periodic factors (e.g., lunar effects) must be considered in chamber-volume growth.

The sequence of chambers and their volumes characterizes cell growth of the foraminifer. Well-preserved tests can, thus, be used for determining growth and lifetime of foraminifera not only for individuals living in the natural environment, but also in historic and fossil forms using the actualistic approach.

In the following, methods for estimating growth pattern and lifetime in multi-chambered foraminifera and the detection of deviations from constant growth will be demonstrated. These methods are performed on living symbiont-bearing benthic foraminifera, but can be used for almost all multi-chambered foraminifera. Because these methods are restricted to volumes, they are independent of chamber form and shell form.

3.2 Growth Functions

As can be found in many textbooks (e.g. Batschelet 1971), growth in biology—either in organisms or in populations and clones—can be fitted by two mathematical models: the exponential growth model and the logistic growth model.

The *exponential growth* following

$$f(t) = a \exp(rt) \tag{3.1}$$

(where r is the growth rate, t determines time and a is the initial value at $t=0$), is unlimited.

The *logistic growth* is restricted by the capacity factor K, thus following the equation

$$f(t) = \frac{K}{1 + a \exp(-rt)} \tag{3.2}$$

Two modifications are applied to these original models to better approximate natural growth. These are the generalized logistic function and the Gompertz function.

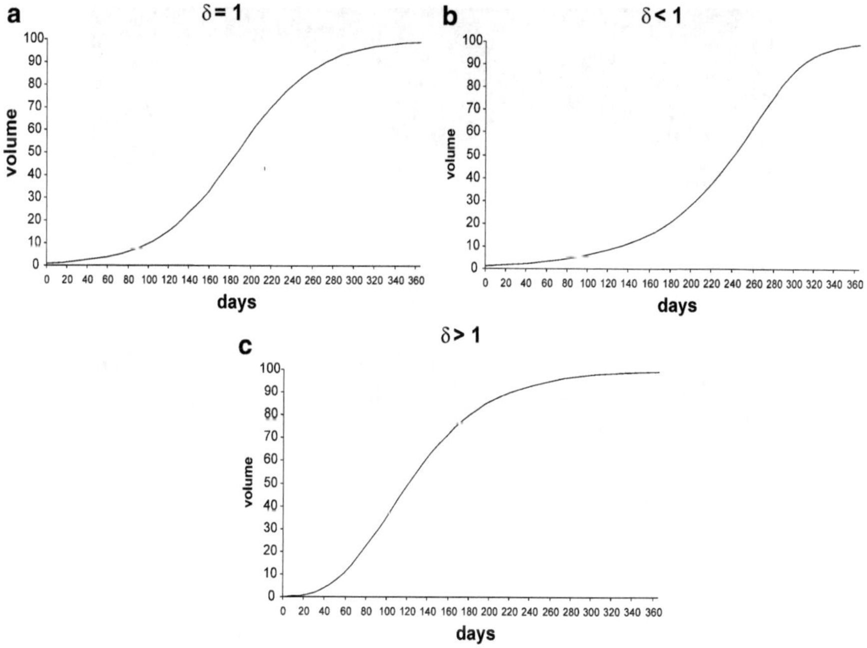

Fig. 3.1 The generalized logistic growth function with different δ-values. (**a**) The normal (symmetrical) logistic function, (**b**) slow initial increase, strong final diminution, (**c**) strong initial increase, slow final diminution

The *generalization* of the *logistic function* (Richards 1959) introduces parameter δ, which determines the differences in the increase and decrease of growth rates of the logistic function

$$f(t) = \frac{K}{\left(1 + \exp\left(-(\lambda + \beta t)/\delta\right)\right)^{\delta}} \tag{3.3}$$

Parameter λ corresponds indirectly to the initial value a, and parameter β to growth rate r in the normal logistic function. Values of $\delta < 1$ marks a weak initial increase and a strong decrease close to the capacity limit, while $\delta > 1$ show a strong initial increase followed by a weak decrease (Fig. 3.1).

The *Gompertz function***,**

$$f(t) = K \exp\left(b \exp(ct)\right), \tag{3.4}$$

as an extension of the exponential growth function approximates to an upper asymptote K, where b as the time-displacement and c as the growth rate are negative numbers. The Gompertz function can also be derived from the generalized logistic function.

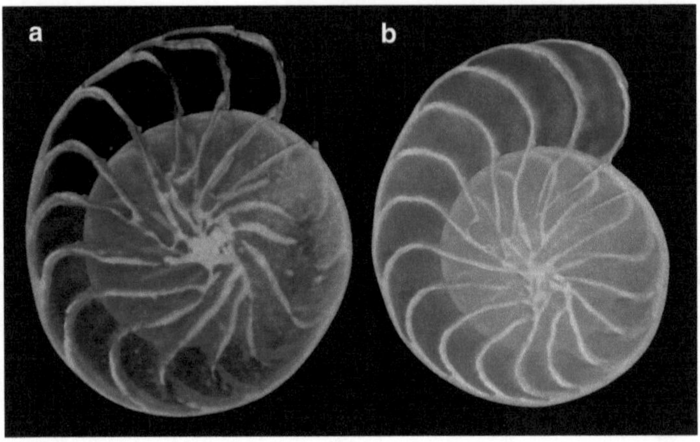

Fig. 3.2 MicroComputerTomography pictures of *Palaeonummulites venosus* from Belau. (**a**) Individual V2 (diameter 3.11 mm), (**b**) individual V3 for comparison (diameter 3.09 mm)

Since in foraminifera the chamber-building rate is primarily unknown and must be estimated by culturing methods or under natural conditions, the chamber number j as a time dependent variable replaces time t in Eqs. (3.1)–(3.4). This transforms the independent variable from a continuous variable t

$$t \in \mathrm{R}_0^+ \left(\textit{the set of positive real numbers}\right) \textit{with } 0 \le t \le \infty$$

to the discrete variable j

$$j \in \mathrm{N}\left(\textit{the set of natural numbers}\right) \textit{with } j = \{1, 2, 3, \ldots, m\}$$

Using time t as the independent variable, $t=0$ marks the reproductive event, while $j=1$ determines the first chamber, the proloculus or protoconch, when chamber number represents the independent variable replacing time.

The growth of shell volume as a function of chamber number $V=f(j)$ can now be modelled using growth functions (3.1)–(3.4).

All methods will be demonstrated using specimens of *Palaeonummulites venosus* (Fichtel and Moll) collected from 50m depth East of Babeldoap, Belau (Hohenegger 1996) as an example (Fig. 3.2a).

The **function tables** for the observed chamber volumes v_{oj} and the observed shell volumes

$$V_{ok} = \sum_{j=1}^{j=k} v_{oj} \quad 1 \le k \le m \tag{3.5}$$

are shown in Table 3.1.

Table 3.1 Chamber volume measurements of *Palaeonummulites venosus*

Chamber number	Day	Chamber volume in mm³	Shell volume in mm³	Cubic root of chamber volume	Shell volume
1		9.80E-05	9.80E-05	0.046	0.046
2		3.40E-05	1.32E-04	0.032	0.079
3		3.00E-06	1.35E-04	0.014	0.093
4	1.2	2.10E-05	1.56E-04	0.028	0.121
5	2.4	3.50E-05	1.91E-04	0.033	0.153
6	3.7	4.20E-05	2.33E-04	0.035	0.188
7	5.0	3.50E-05	2.68E-04	0.033	0.221
8	6.3	6.00E-05	3.28E-04	0.039	0.260
9	7.7	3.10E-05	3.59E-04	0.031	0.291
10	9.1	6.00E-05	4.19E-04	0.039	0.330
11	10.5	1.42E-04	5.61E-04	0.052	0.383
12	12.0	1.67E-04	7.28E-04	0.055	0.438
13	13.5	1.97E-04	9.25E-04	0.058	0.496
14	15.1	4.45E-04	1.37E-03	0.076	0.572
15	16.7	5.91E-04	1.96E-03	0.084	0.656
16	18.4	5.72E-04	2.53E-03	0.083	0.739
17	20.1	1.84E-03	4.37E-03	0.122	0.862
18	21.9	1.09E-03	5.46E-03	0.103	0.964
19	23.7	1.25E-03	6.71E-03	0.108	1.072
20	25.6	1.98E-03	8.69E-03	0.126	1.198
21	27.5	2.52E-03	1.12E-02	0.136	1.334
22	29.6	3.02E-03	1.42E-02	0.145	1.478
23	31.6	3.66E-03	1.79E-02	0.154	1.632
24	33.8	2.24E-03	2.01E-02	0.131	1.763
25	36.0	5.57E-03	2.57E-02	0.177	1.941
26	38.3	6.25E-03	3.19E-02	0.184	2.125
27	40.7	5.78E-03	3.77E-02	0.179	2.304
28	43.2	8.13E-03	4.58E-02	0.201	2.505
29	45.8	7.28E-03	5.31E-02	0.194	2.699
30	48.5	6.27E-03	5.94E-02	0.184	2.883
31	51.3	3.87E-03	6.33E-02	0.157	3.040
32	54.1	4.47E-03	6.77E-02	0.165	3.205
33	57.2	1.30E-02	8.08E-02	0.235	3.440
34	60.3	1.71E-02	9.78E-02	0.257	3.698
35	63.6	1.64E-02	1.14E-01	0.254	3.952
36	67.0	1.61E-02	1.30E-01	0.253	4.205
37	70.5	2.36E-02	1.54E-01	0.287	4.492
38	74.3	2.23E-02	1.76E-01	0.281	4.773
39	78.2	1.78E-02	1.94E-01	0.261	5.034
40	82.3	2.02E-02	2.14E-01	0.272	5.306
41	86.6	2.68E-02	2.41E-01	0.299	5.606
42	91.1	2.65E-02	2.68E-01	0.298	5.904
43	95.8	3.35E-02	3.01E-01	0.322	6.226
44	100.8	3.28E-02	3.34E-01	0.320	6.547
45	106.0	2.51E-02	3.59E-01	0.293	6.839

(continued)

Table 3.1 (continued)

Chamber number	Day	Chamber volume in mm³	Shell volume in mm³	Cubic root of chamber volume	Shell volume
46	111.6	3.06E-02	3.90E-01	0.313	7.152
47	117.5	4.01E-02	4.30E-01	0.342	7.495
48	123.7	4.31E-02	4.73E-01	0.350	7.845
49	130.3	3.56E-02	5.08E-01	0.329	8.174
50	137.3	4.26E-02	5.51E-01	0.349	8.523
51	144.7	3.59E-02	5.87E-01	0.330	8.853
52	152.7	6.19E-02	6.49E-01	0.396	9.249
53	161.2	5.11E-02	7.00E-01	0.371	9.620
54	170.3	5.13E-02	7.51E-01	0.371	9.991
55	180.1	5.87E-02	8.10E-01	0.389	10.380

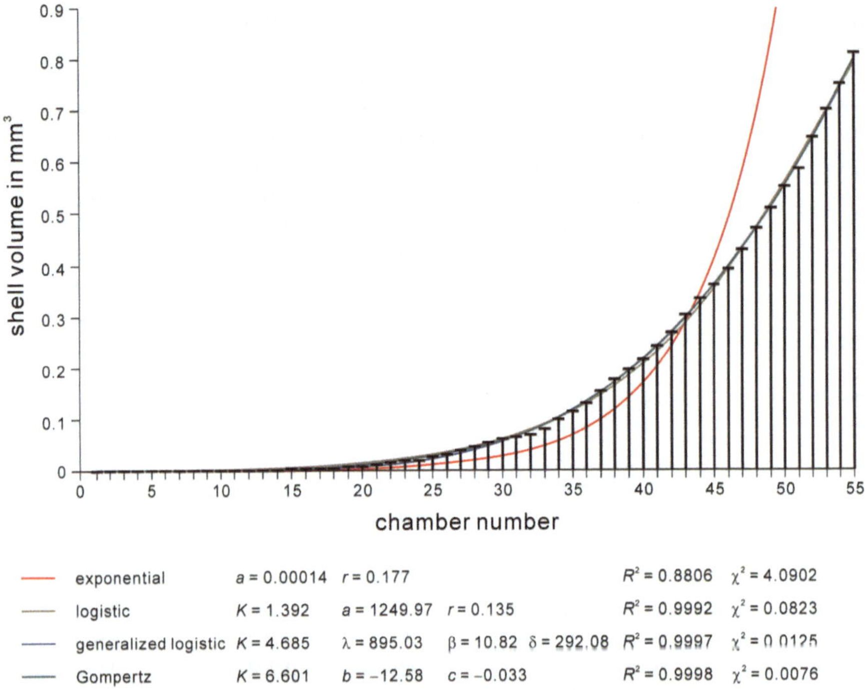

Fig. 3.3 Cell-volume growth of *Palaeonummulites venosus* represented as the sum of chamber volumes. Fit by growth functions [Eqs. (3.1)–(3.4)] with their parameters; quality of fit determined by the 'coefficient of determination' R^2 and the χ^2-statistic

The correct *function graph* of volume growth using chamber number j as the independent variable is the *bar diagram* (Fig. 3.3), because intervals between integers have no meaning (a histogram representing mean frequencies in intervals of the number line is the correct function graph for continuous variables, e.g. time divided into intervals like weeks, years etc.).

The quality of fit can be proven in two ways. The *coefficient of determination* R^2 that is found in all statistical programs for determining correlation explains the grade of similarities in the function form between observed o and expected e values of the dependent variable y (in our case shell volumes V_o, V_e and chamber volumes v_o, v_e) from the independent variable x (in our case chamber number j)

$$R^2 = \left[\sum_{i=1}^{i=n} \left(y_{oi} - \bar{y}_o \right) \left(y_{ei} - \bar{y}_e \right) \right]^2 / \left[\sum_{i=1}^{i=n} \left(y_{oi} - \bar{y}_o \right)^2 \sum_{i=1}^{i=n} \left(y_{ei} - \bar{y}_e \right)^2 \right] \tag{3.6}$$

The intensity of deviation of the observed values from the expected function can be measured by the *Chi-square statistic*

$$\chi^2 = \sum_{i=1}^{i=n} \frac{\left(y_{oi} - y_{ei} \right)^2}{y_{ei}} \tag{3.7}$$

In this case, it can be demonstrated that the best fit of shell volume growth is by the Gompertz function, closely followed by the generalized logistic function (Fig. 3.3).

Using chamber number j as the independent variable replacing time, the expected chamber volumes v_e according to growth functions calculated with Eqs. (3.1)–(3.4) can be calculated using their first derivatives $v_e = V_e' = dV_e/dj$ leading to

$$\frac{dV_e}{dj} = rV_e, \text{ which is } \textit{exponential growth}; \tag{3.8}$$

$$\frac{dV_e}{dj} = \frac{r}{K} V_e \left(K - V_e \right), \text{ which is } \textit{logistic growth}; \tag{3.9}$$

$$\frac{dV_e}{dj} = \beta V_e \left(1 - \left(V_e / K \right)^{1/\delta} \right), \text{ which is } \textit{generalized logistic growth}; \tag{3.10}$$

and $$\frac{dV_e}{dj} = cV_e \ln \left(K / V_e \right), \text{ which is the } \textit{Gompertz function} \tag{3.11}$$

Except for differential Eq. (3.8), Eqs. (3.9)–(3.11) are quadratic functions that determine the maximum increase at $V_e' = 0$ characterizing the inflection point. Starting from this point, chamber volumes decrease during further growth.

In foraminifera, the *inflection point* marks the time t or chamber number j, when under undisturbed environmental conditions reproduction typically occurs. Further growth leads to smaller chambers, often found in planktic foraminifera (Olsson 1973), which are called 'kummerformen'.

Chamber volumes and fit by the first derivative of the growth functions are shown in Fig. 3.4. Again, the Gompertz function shows the best fit.

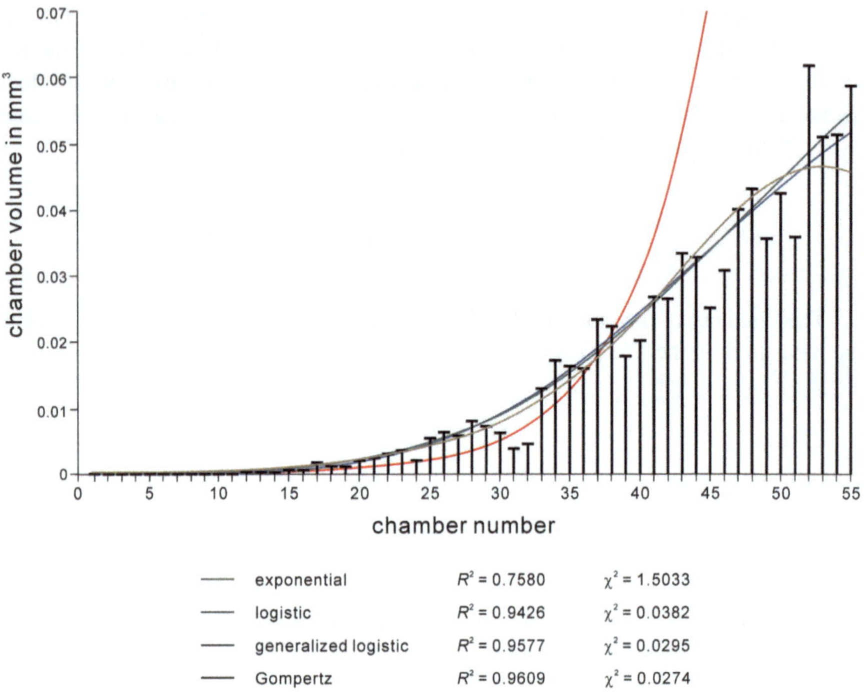

Fig. 3.4 Chamber-volume growth of *Palaeonummulites venosus*. Fit by first derivatives [Eqs. (3.8)–(3.11)] based on parameters of function (3.1)–(3.4); quality of fit determined by the 'coefficient of determination' R^2 and χ^2-statistic

3.3 Residuals

Periodic or instantaneous deviations from the growth function can be obtained by calculating *residuals d*. In the generalized function $y=f(x)$ these are the differences between the observed values y_{oi} dependent on x_i and the expected values y_{ei} given by the growth function, thus

$$d_i = y_{oi} - y_{ei} \tag{3.12}$$

where i determines the series of measurements ($i=1, 2, 3, \ldots, n$).

Residuals can be calculated for shell growth V and chamber volume growth v, both replacing the dependent variable y in Eq. (3.12), where i is replaced by j (Fig. 3.5).

Residuals increase significantly with chamber number in both shell volume and chamber volumes growth (Fig. 3.5). This is caused by the non-linearity of volumes that increase drastically (3rd power) with time. Therefore, deviations from the regression function at early stages are much smaller than deviations in later growth stages. This problem can be solved calculating *standardized residuals* by

$$d_i = \left(y_{oi} - y_{ei}\right) / \, y_{ei} \tag{3.13}$$

relating deviations to the expected values given by the growth functions (Fig. 3.6).

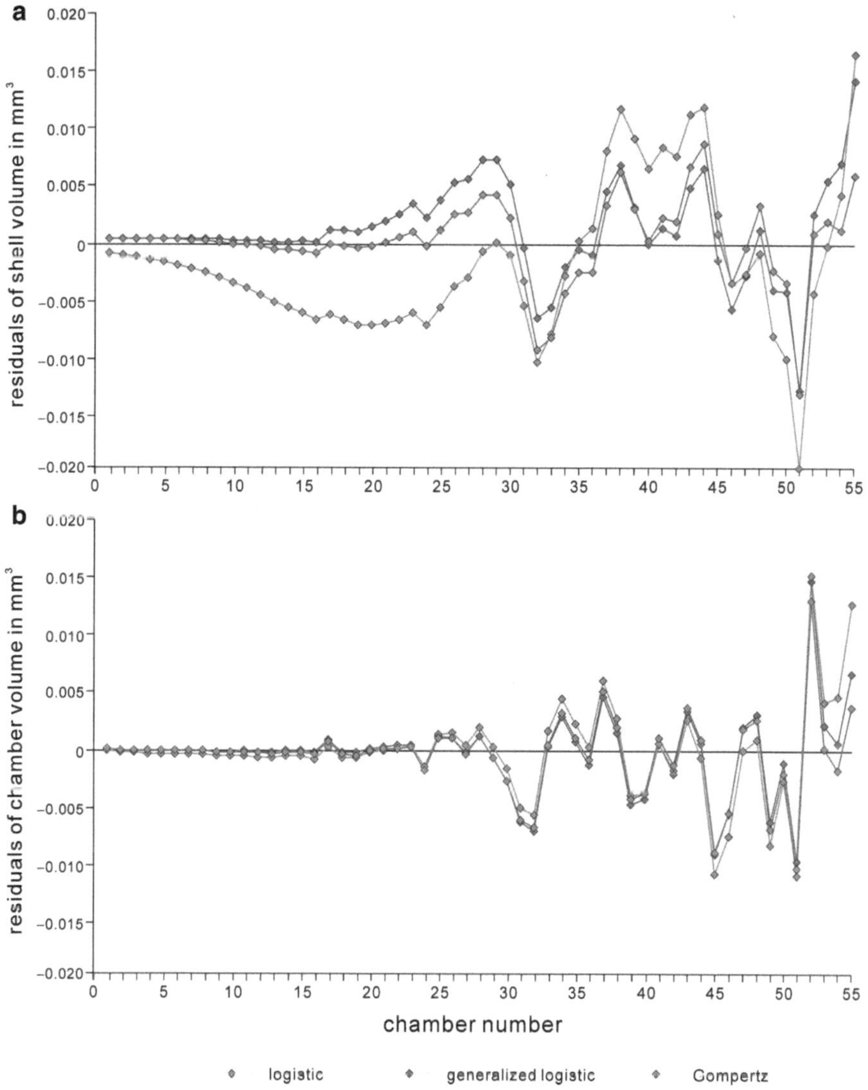

Fig. 3.5 Residuals of growth functions. (**a**) shell volume, (**b**) chamber volumes

In contrast to normal residuals, standardized residuals significantly decrease from early to late stages in both volume measures, where the decrease in shell volume is much stronger than in chamber volumes (Fig. 3.6). Nevertheless, periodicities in deviations from the mean growth function can now be clearly observed using standardized residuals of chamber volumes.

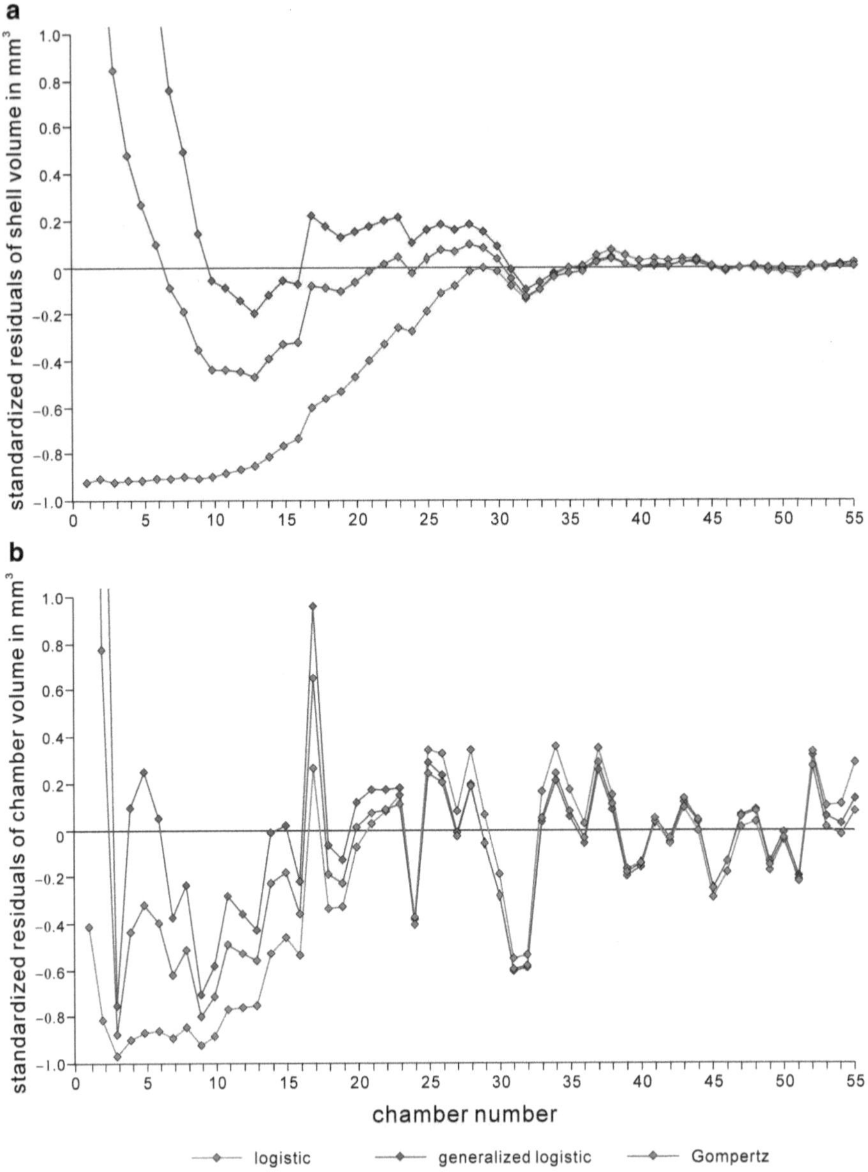

Fig. 3.6 Standardized residuals of growth functions. (**a**) shell volume, (**b**) chamber volumes

3.4 Transformation of Non-linear Data

The above mentioned difficulties in linear statistical methods using nonlinear variables like volumes can be resolved by transformation of the dependent variable to linear data. *Linearization* is easy and can be achieved by calculating the cubic root

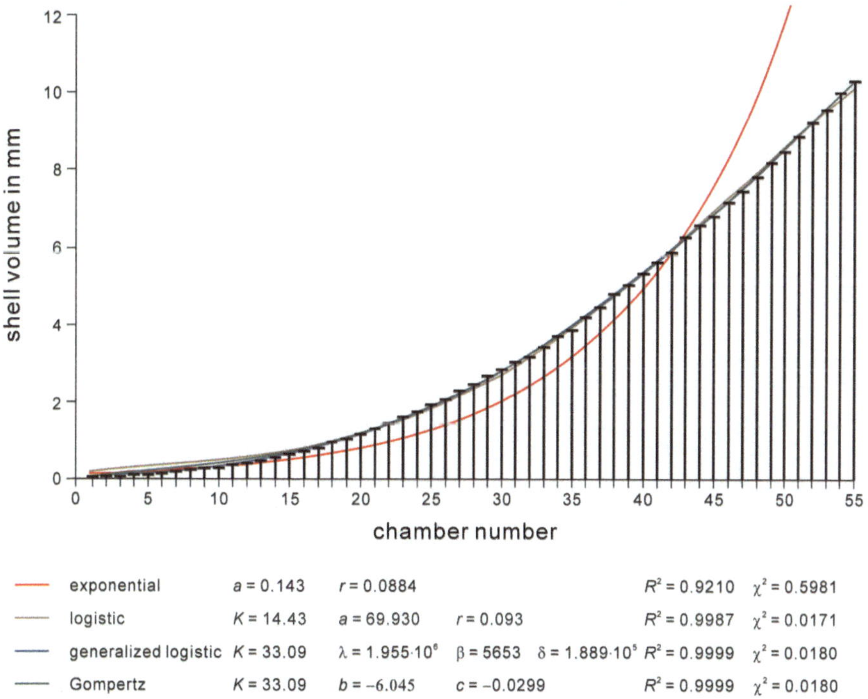

Fig. 3.7 Cell-volume growth of *Palaeonummulites venosus* represented as the sum of linearized chamber volumes. Fit by growth functions [Eqs. (3.1)–(3.4)] with their parameters; quality of fit determined by the 'coefficient of determination' R^2 and the χ^2-statistic

of chamber volumes leading to

$$V_{ok}^{*} = \sum_{j=1}^{j=k} \left(v_{oj} \right)^{1/3} \text{ when } 1 \le k \le m \tag{3.14}$$

The fit of the linearized shell volume by growth function Eqs. (3.1)–(3.4) shows a much better fit compared to the non-transformed volumes in all functions. In linearized shell volumes, the fit by the generalized logistic function and the Gompertz function are identical (Fig. 3.7).

Using differential Eqs. (3.8)–(3.11), the growth of transformed chamber volumes can be calculated (Fig. 3.8).

For determining intensities of deviations from theoretical growth functions, residuals of the observed cubic roots of chamber volumes to the expected chamber volumes obtained by the first derivatives of the growth functions can be calculated.

The increase in absolute values of standardized residuals with chamber number is significant in all functions (Fig. 3.9a), but much weaker compared to non-linearized chamber volumes (Fig. 3.5b). The positive deviation of the first two

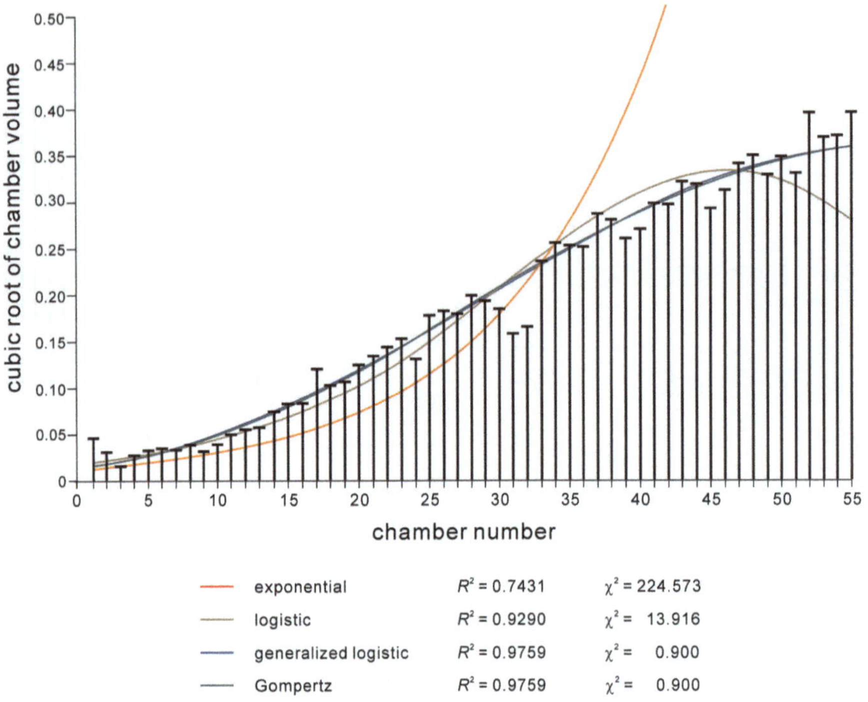

Fig. 3.8 Growth of cubic roots of chamber volumes in *Palaeonummulites venosus*. Fit by first derivatives [Eqs. (3.8)–(3.11)] based on parameters of function (3.1)–(3.4) (Fig. 3.7), quality of fit determined by the 'coefficient of determination' R^2 and χ^2-statistic

chambers in *P. venosus* from the growth functions is typical for nummulitids, where the proloculus and deuteroloculus characterizing the embryonic apparatus are much larger than the succeeding chambers.

Standardized residuals of linearized chamber volumes again demonstrate a significant decrease with chamber number (Fig. 3.9b), but are much weaker compared to non-linearized chamber volumes (Fig. 3.6b). Deviations in the earlier shell part are much stronger compared to the later part. The protoconch and deuteroconch in particular show extreme residuals.

Oscillations are clearly demonstrated in normal and standardized residuals of chamber volumes and the cubic root of chamber volumes (Figs. 3.5b, 3.6b, and 3.9a, b). The position of peaks and valleys are identical; the differences are expressed in the amplitudes. As explained above, amplitudes are small in the initial part of the shell increasing towards the final chambers using non-standardized residuals (Figs. 3.5b, 3.9a), while the opposite trend—large amplitudes at the beginning and smaller amplitudes at the end of shell construction—are typical for standardized residuals (Figs. 3.6b, 3.9b).

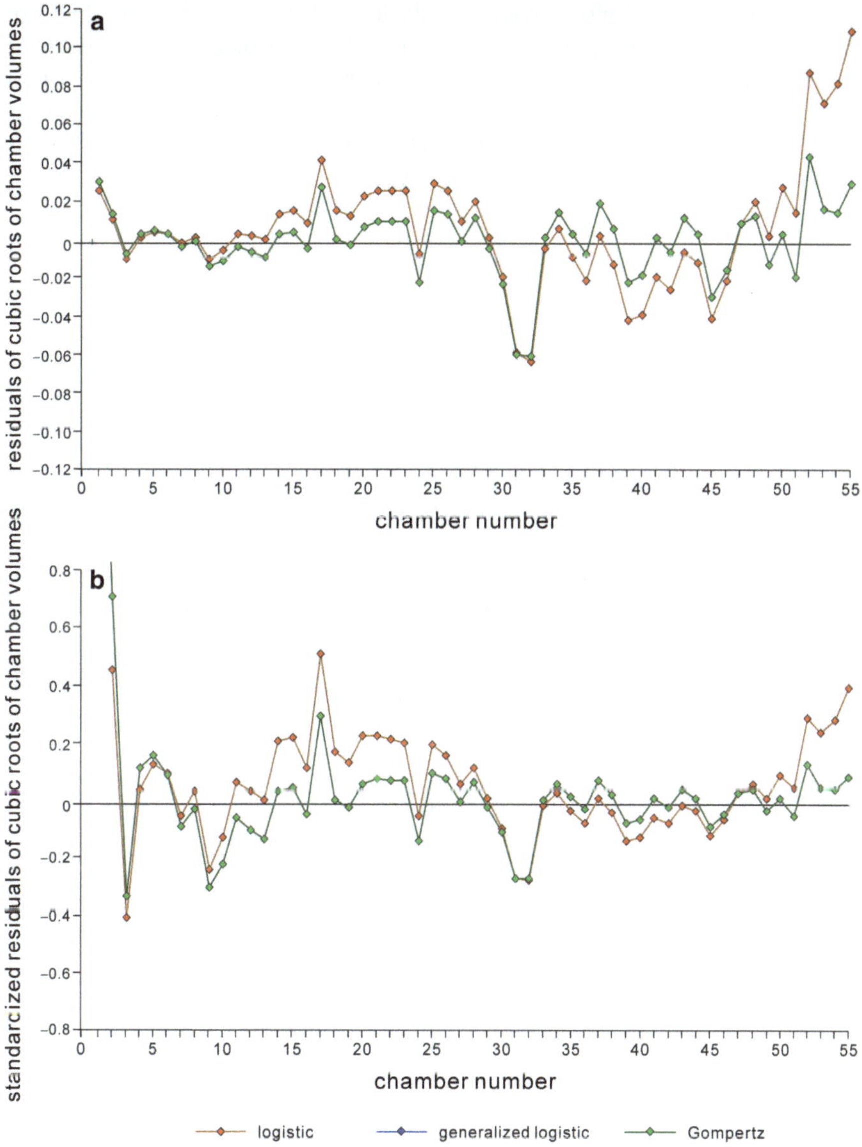

Fig. 3.9 Residuals on growth functions using cubic roots of chamber volumes. (**a**) Normal residuals, (**b**) standardized residuals. Mark identical residuals using generalized logistic growth and the Gompertz function

3.5 Comparing Oscillations

Similarities between individuals in deviations and oscillations from the mean growth function can be tested with the method of *cross-correlation*.

One requirement of this method is that the intervals (= lags) between values y on the independent variable x must be constant. If the independent variable is time t, then constant intervals can be days (lag = 1), weeks (lag = 7 days) or years, but not calendar months, which vary in length. Because intervals in the above example are chamber numbers replacing time, the comparison by cross-correlation is possible because of constant interval steps (lag = 1 chamber).

Cross-correlation is defined by

$$r_n = \frac{n^*\left(\sum y_{i1}y_{i2}\right) - \sum y_{i1}\sum y_{i2}}{\sqrt{\left[n^*\sum y_{i1}^2 - \left(\sum y_{i1}\right)^2\right]\left[n^*\sum y_{i2}^2 - \left(\sum y_{i2}\right)^2\right]}} \tag{3.15}$$

where n^* marks the number of overlapping observations. The first sequence $y_{(i)\,1}$ with n_1 observations, where brackets determine the time order within the series of observations (David 1970), will be kept stable, while the second sequence $y_{(i)\,2}$ with observations n_2 must be scattered in steps (= lags) along the first sequence. Scattering starts at least at $y_{(1)\,1}$ together with $y_{(n_2-3)2}$, decreasing n_2 step by step and finishing at least when $y_{(1)\,2}$ coincides by shifting with $y_{(n_1-3)1}$. Smaller intervals are possible because coincidences with few overlapping observations, as can be found at both ends of the lag scale, are often highly significant (Fig. 3.10).

The significance of correlation must be checked for each lag by testing the probability of non-correlation using the t-distribution of

$$t_0 = r_{n^*}\sqrt{\frac{n^* - 2}{1 - r_{n^*}^2}} \tag{3.16}$$

The best concordance is performed by lags (= number of shifts) with positive correlation and low probability of non-correlation (Fig. 3.11), but only lags close to lag 0 are relevant, because overlapping periods at both ends of the lag scale are extremely short and cannot characterize the coincidence of periods (e.g., lag −32 and lag +31 in Fig. 3.10).

Overlapping and identical periods can only be measured when using the chamber number for obtaining lags. Due to the different time intervals between the construction of chambers, these oscillations cannot directly be transferred into time, especially when the significant lag closest to 0 is too far away, because the chamber-building rate extends the time intervals during growth, making the comparison of residuals impossible using identical chamber numbers for time.

3.6 Chamber Building Rate

For the detection and explanation of time-dependent oscillations in chamber volumes, the independent variable chamber number j has to be transferred into time t. The function $j = f(t)$ is called the *chamber-building rate*.

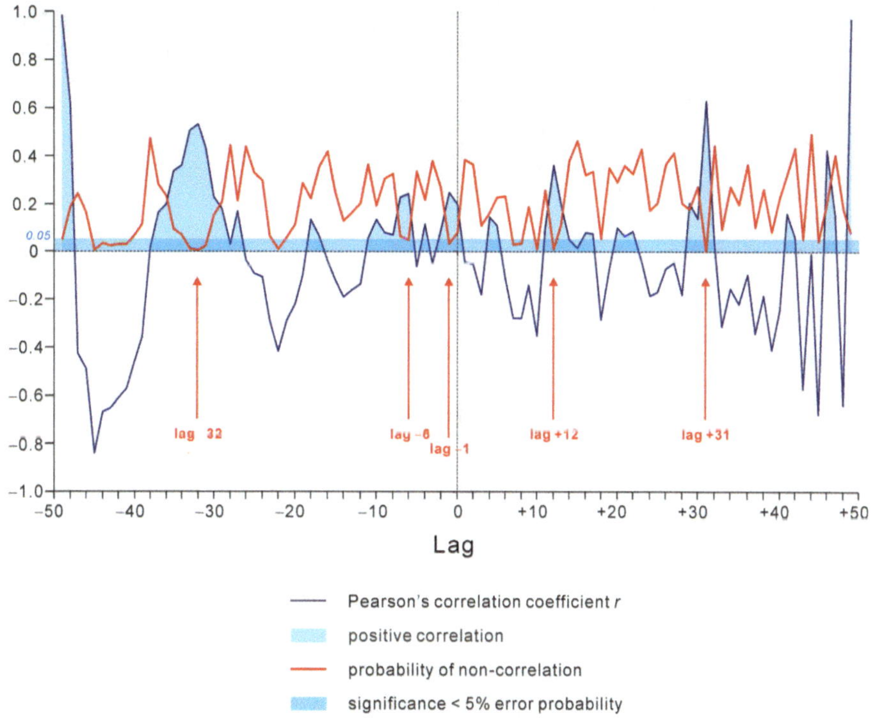

Fig. 3.10 Cross-correlation between 2 individuals of *Paleonummulites venosus* (samples V2 and V3) with $n_1=55$ chambers and $n_2=55$ chambers based on standardized residuals. Significant correspondence between oscillations are marked by positive correlations, when the probability of non-correlation sinks below 0.05 (=5 % error probability)

Investigations of chamber construction in nummulitids have been done under laboratory conditions only for *Heterostegina depressa* (Röttger 1972; Röttger and Spindler 1976) and *Cycloclypeus carpenteri* (Lietz 1996), where the former investigation was successful after an instantaneous change in light conditions to 300 lux, which is typical of in situ conditions. Under these conditions, the time of chamber construction could be measured for 20 specimens of *H. depressa*.

A good fit of the chamber-building rate is by the *Michaelis–Menten* function

$$j = f(t) = (at)/(b+t) \tag{3.17}$$

where a is the rate of increase and b determines the diminution factor at time t (Fig. 3.12).

In the following, the averaged Michaelis–Menten function was calculated for the chamber building rate of 20 specimens of *Heterostegina depressa* under laboratory conditions (Fig. 3.13).

For checking the influence of environmental effects, the chamber-building rate must be calculated under natural conditions. Knowing the onset of reproduction, it can be used for comparing additional *H. depressa* populations.

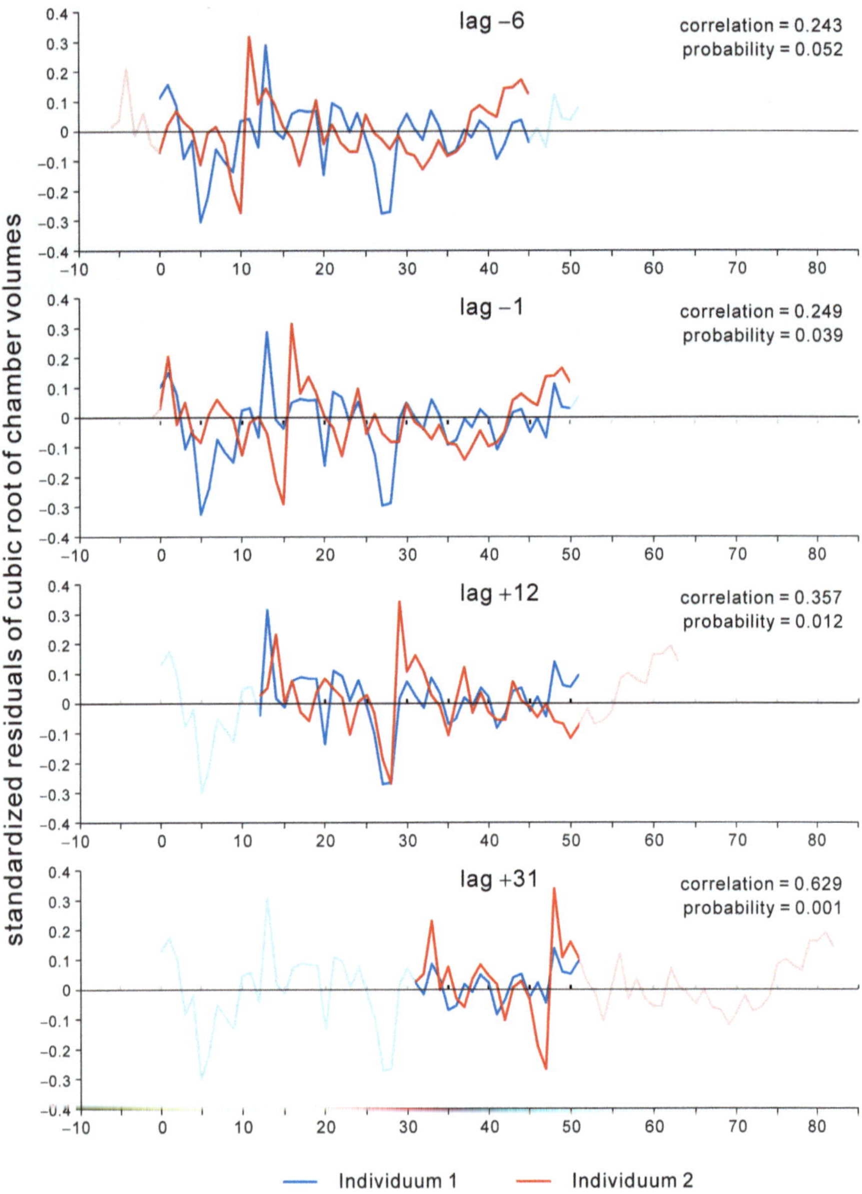

Fig. 3.11 Standardized residuals of 2 individuals of *Paleonummulites venosus* at significant lags obtained by cross-correlation (Fig. 3.10). Note the short interval of significant overlapping at lag +31 that is close to the limits of the lag scale (lag –49 and lag +49)

Using time as the independent variable, growth values of the shell and chamber volumes obtained by chamber number as the independent variable remain stable, while chamber number j must be replaced by time t (measured in days) following

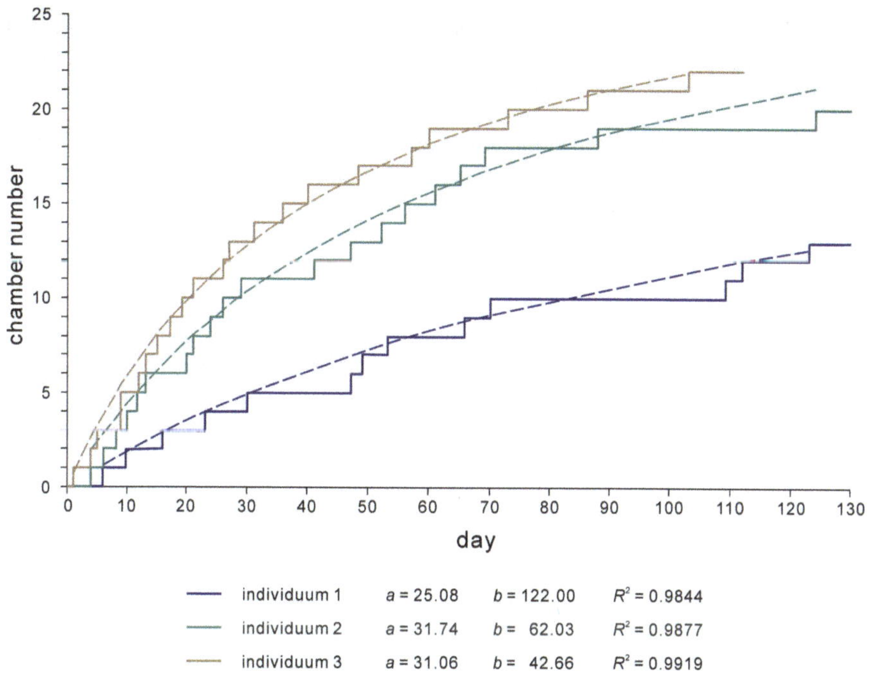

Fig. 3.12 Chamber building of 3 *Heterostegina depressa* specimens under laboratory conditions and fit by the Michaelis–Menten function

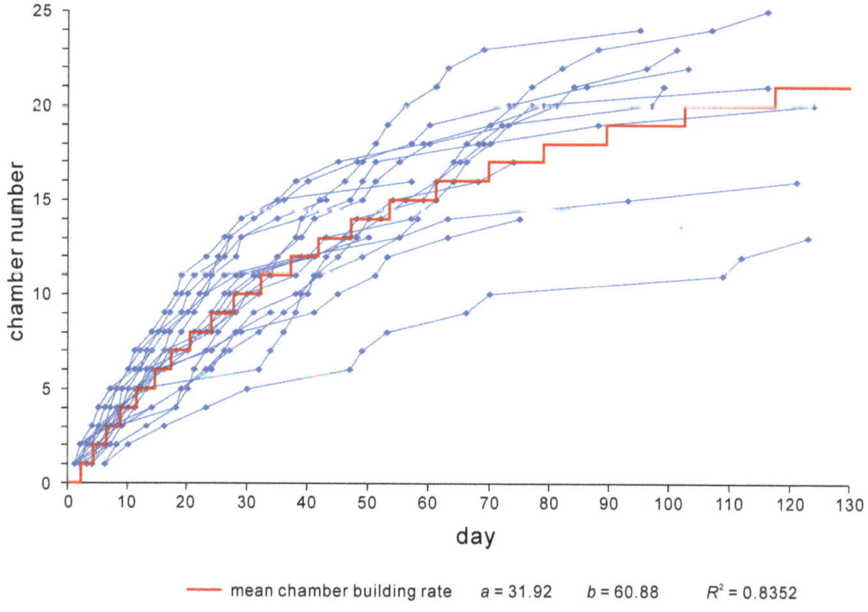

Fig. 3.13 Chamber building of 20 *Heterostegina depressa* specimens under laboratory conditions and fit by the averaged Michaelis–Menten function

$$t = (bj)/(a - j) \tag{3.18}$$

using the parameter values of Eq. (3.15).

Because in all investigated nummulitids, the offspring from the agamont or schizont consist of 3 chambers, chamber construction dependent on time starts with the 4th chamber. Therefore, the chamber-building rate for nummulitids transforms to

$$t = b(j - 3)/(a - j + 3) \tag{3.19}$$

Based on the initial chamber buildings experienced by Krüger (1994) in the laboratory but estimating the construction of the last chamber in our example of *P. venosus* at day 180 under natural conditions, the intervals in days between the *j* chambers can be calculated by

$$t = 91.192(j - 3)/(78.912 - j + 3)$$

Therefore, shell-volume growth was calculated related to time, where the function graph is now a stepwise function because time is a continuous variable (Fig. 3.14a). The cubic roots of chamber volumes still must be represented as bar diagrams (Fig. 3.14b).

Intervals between the standardized residuals of chamber volumes or between the non-standardized and standardized residuals of the cubic root of chamber volumes can be calculated, allowing the interpretation of time-dependent deviations in all residuals (Fig. 3.15).

Cross-correlation testing the coincidence of periods between the individuals cannot be used directly because of the inconstant intervals as mentioned above. But the significance of oscillations can be checked using time-series analyses that are independent of constant intervals between observations.

3.7 Further Investigations

Several consecutive investigations can be based on volume growth indicating environmental influence, especially differentiating instantaneous from periodic deviations. Because the natural chamber-building rate in the above samples was not tested and lifetime was assumed to be 180 days for *P. venosus*, the following methods are mentioned for thoroughness but not explained in detail.

3.7.1 Determining Outliers

Instantaneous deviations caused by unfavorable short-time environmental conditions (e.g., non-lethal attack by predators) are expressed in spontaneous strong negative excursions of standardized residuals. These can be tested using the

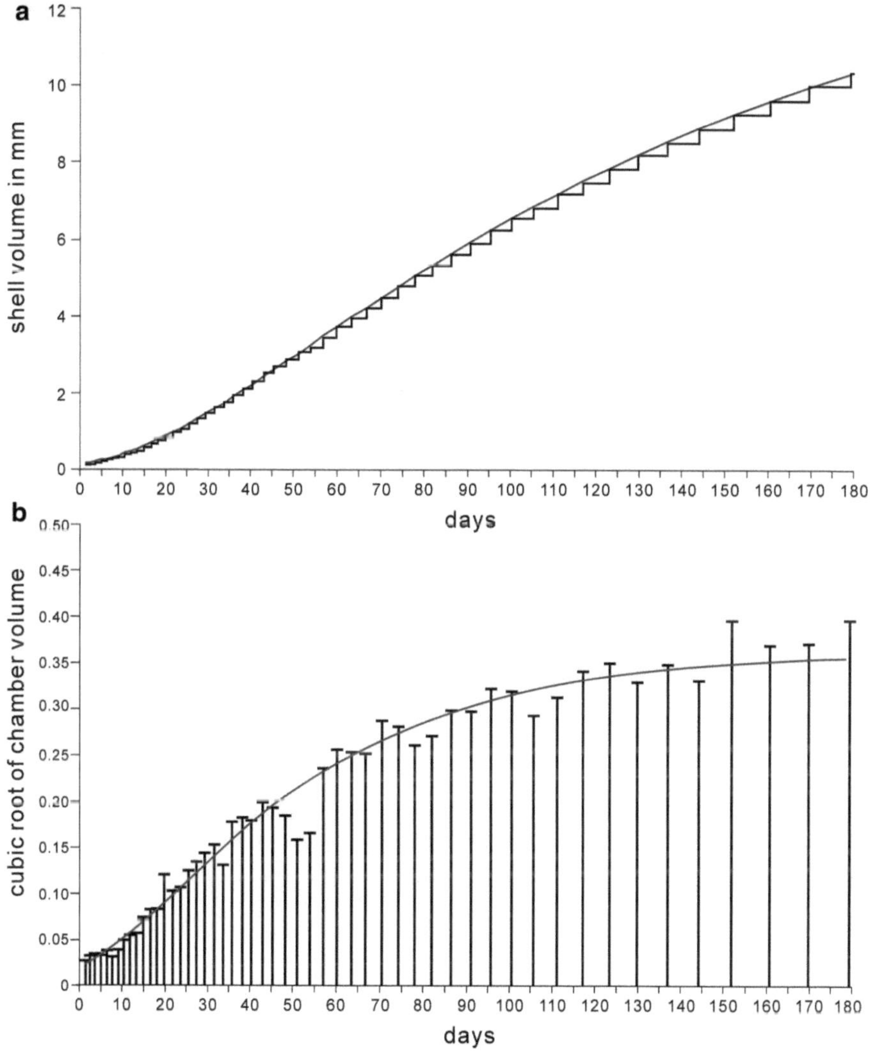

Fig. 3.14 (**a**) Cell-volume growth of *Palaeonummulites venosus* represented as the sum of linearized chamber volumes dependent on time and fit by Gompertz function based on chamber number (Fig. 3.7), (**b**) Growth of cubic roots of chamber volumes and fit by the first derivative of the Gompertz function based on chamber number (Fig. 3.8)

frequency distribution of standardized residuals. Calculating the mean and standard deviation of residuals, outliers that do not belong to the given normal distribution can be detected by the position of a specified quantile (or percentile), where the probabilities belonging to the given distribution are lower than an error limit of 10 %, 5 % or 1 % probability.

Negative residuals lying outside the error limits must not automatically point to short-term events. When the negative outlier is followed by a positive outlier or values close to the upper error limit then both outliers are *balanced* indicating strong oscillation (Fig. 3.16).

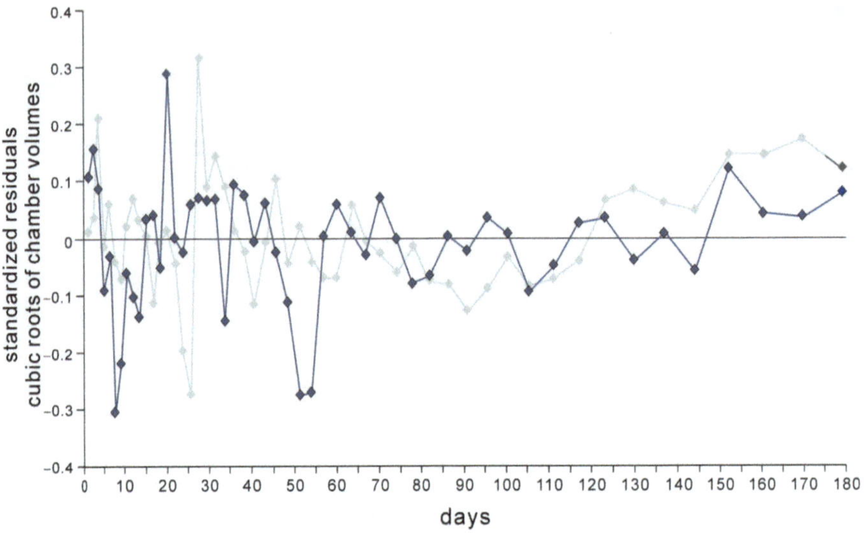

Fig. 3.15 Standardized residuals of the cubic root of chamber volumes fitted by Gompertz function in relation to time; residuals of the second individual in light color

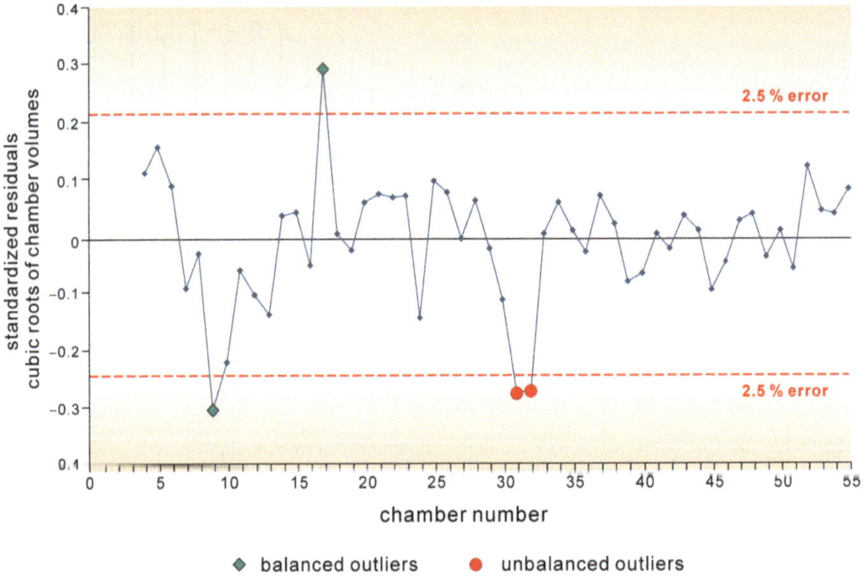

Fig. 3.16 Standardized residuals of cubic root of chamber volumes fitted by Gompertz function. The position of 5 % error limits and determination of outliers belonging to oscillating deviations (balanced outliers) and to spontaneous deviations (unbalanced outliers)

Strong negative oscillations without positive counterparts, so-called *unbalanced* outliers, point to deviations caused by instantaneous events, sometimes followed by recovery (Figs. 3.16, 3.17).

The initial chambers, i.e., the protoconch, deuteroconch and the 3rd chamber belong to the embryonic apparatus of nummulitids. They are constructed within the parent shell

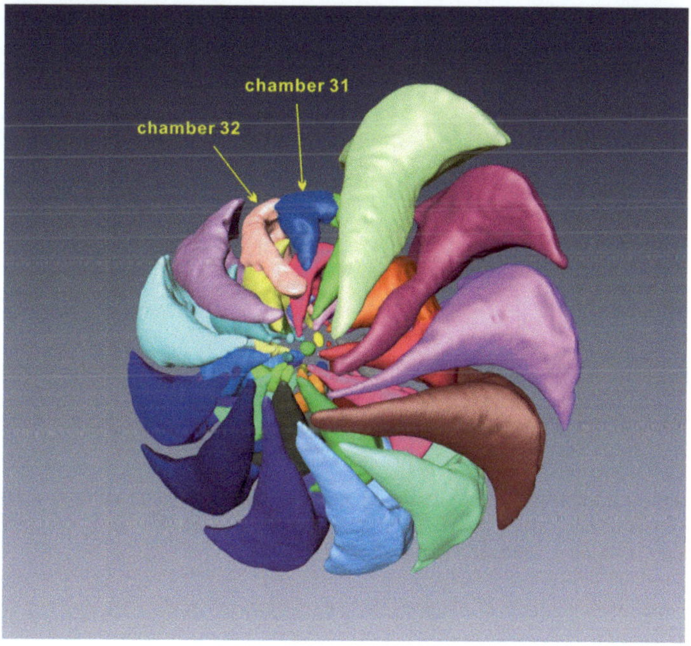

Fig. 3.17 MicroCT presentation of the first 42 chamber volumes of *Palaeonummulites venosus* demonstrating that chambers 31 and 32 are unbalanced outliers with significantly smaller volumes

and thus protected from environmental influences until offspring emerge from the parental test/shell. Deviations from the mean chamber volume growth function of these first chambers are extreme because the protoconch and deuteroconch are much larger than the following chambers (Fig. 3.9b). Therefore, these outliers are inherent to the system.

3.7.2 Time-Series Analysis

Periodic oscillations should be tested by statistical methods derived from Fourier analyses (Batschelet 1971). Complex oscillations can be defined as the sum of k sinusoidal oscillations following

$$y = y_0 + \sum_{j=1}^{j=k} a_j \cdot \cos\left(\frac{2\pi}{b_j} \cdot t - c_j\right) \tag{3.20}$$

where a_j indicates the amplitude, b_j indicates the period and c_j the phase of the jth sinusoidal function, all oscillating around the constant y_0.

Fourier-Analysis needs equal time intervals, but some *power-spectral analyses* work with inconsistent time intervals, as this is necessary in our example with increasing time intervals due to chamber-building rate.

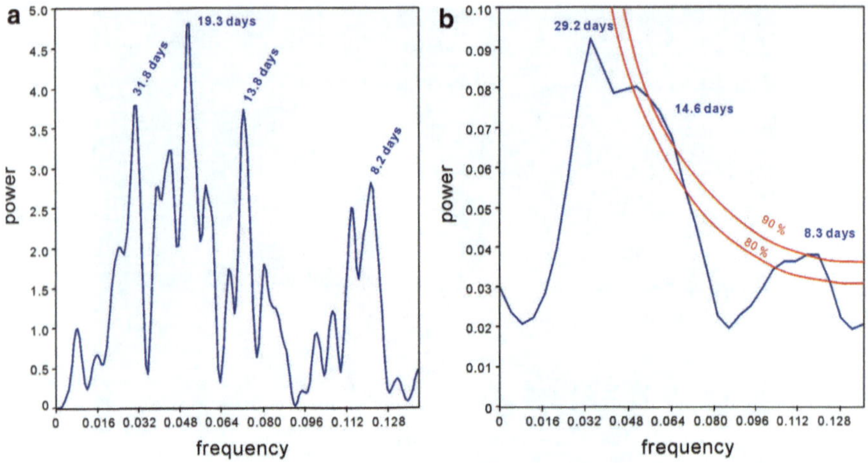

Fig. 3.18 Power spectral analysis. (**a**) Lomb periodogram showing significant periods, (**b**) REDFIT spectral analysis demonstrating the effect of noise reduction in oscillations by averaging overlapping segments. *Red lines* indicate false alarm levels

In power-spectral analyses, significant periods can be detected by decomposing the huge number of single oscillations into a fewer number of sinusoidal oscillations characterized by frequencies with significant or even strong power (Davis 2002). Two methods are shown below to demonstrate the search for significant periods and the possibilities of interpretation.

The first method of power spectral analysis is by *Lomb-periodogram*, which shows the significant periods as power peaks (Fig. 3.18a). Since the data number is low (n=52), the significance of the peaks cannot be reached neither by 1 % error probability (power 9.21) nor by 5 % error probability (power 7.576). Nevertheless, there are high peaks that can be used for calculating compound sinusoidal oscillations. The lacking parameters for the amplitudes (a_j), phases (c_j) and the constant y_0 (should be close to 0) can be estimated using regression analyses (Table 3.2, Fig. 3.19).

The *REDFIT spectral analysis* (Schulz and Mudelsee 2002) splits the time series into a number of segments, overlapping by 50 %, and averaging their spectra. Therefore, the data are smoothed and results concentrated, leading to reduction of significant peaks (e.g., the dominant peak at 19.3 days in the Lomb-periodogram) and slight shifts in the power peaks, explaining their significance through "false alarm levels" (Fig. 3.18b). Again, the lacking parameters of the remaining 3 sinusoidal functions can be calculated by regression analyses (Table 3.2).

An important result in the chamber-building rate is the clear presence of a two-week and a four-week cycle obtained by both power-spectra analyses. The summing of these cycles results in a continuous alternation of a weaker peak followed by a stronger peak (Fig. 3.20). This allows interpreting the influence of spring tides,

Table 3.2 Results of power spectral analyses

Amplitude *a*	Period *b*	Phase *c*
Lomb periodogram		
0.0459	19.34	–0.266
0.0488	31.80	–1.020
0.0381	13.76	0.301
0.0341	8.22	2.000
Constant y_0	0.0021	
Chi-square	0.2343	
R-square	0.4824	
Probability	4.79E-03	
REDFIT spectral analysis		
0.0414	8.34	1.500
0.0309	14.60	0.286
0.0242	29.20	–0.446
Constant y_0	–0.0040	
Chi-square	0.3628	
R-square	0.2004	
Probability	2.27E-01	

where the new-moon tide could represent the weaker and the full-moon tide could represent the stronger peak. The cause and effects must be proven.

It must be noted that these calculations were based on the assumption that the last chambers following Krüger (1994) investigations have been built after 180 days, thus longevity is approximately half a year. Extending this to a one-year life cycle, amplitudes and phases remain constant in power spectral analyses, doubling period length in Table 3.2. This makes interpretation by lunar cycles easier, because the most dominant period in REDFIT analysis is now the 2 week spring-tide cycle, followed by the weaker monthly full-moon cycle and the less significant 2-month cycle.

But these interpretations are only possible when the chamber building rates under natural conditions are known; otherwise the influence of time-dependent factors can strongly be biased, especially in chamber-building rates found in cultures.

3.8 Working Steps

From these results, the steps for estimating cell growth in multi-chambered foraminifera based on shells are proposed:

(1) calculate cubic roots of chamber volumes,
(2) calculate shell volumes using Eq. (3.14) (Fig. 3.7),
(3) calculate the generalized logistic function [Eq. (3.3)] and Gompertz function [Eq. (3.4)] based on chamber number *j*,

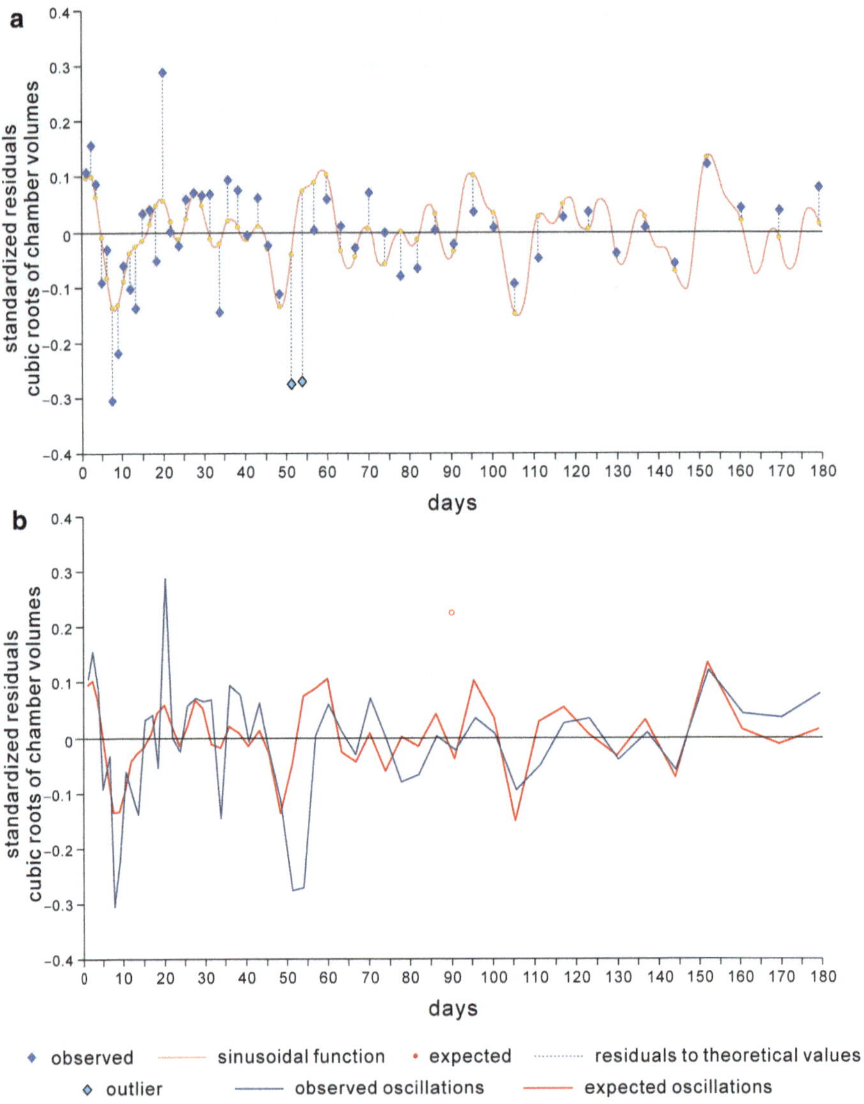

Fig. 3.19 Fit of standardized residuals of cubic roots of chamber volumes by the sum of the four sinusoidal oscillations obtained by power spectral analysis based on Lomb periodograms (Table 3.2, Fig. 3.18a), (**a**) Residuals of empirical data to the composite sinusoidal functions, (**b**) Comparison of observed oscillations with expected composite sinusoidal oscillations

(4) prove the quality of fit by the 'coefficient of determination' [Eq. (3.6)] and 'chi-square statistic' [Eq. (3.7)],
(5) calculate the first derivatives of the generalized logistic function [Eq. (3.10)] and the Gompertz function [Eq. (3.11)] to compare with the observed chamber volumes (Fig. 3.8),

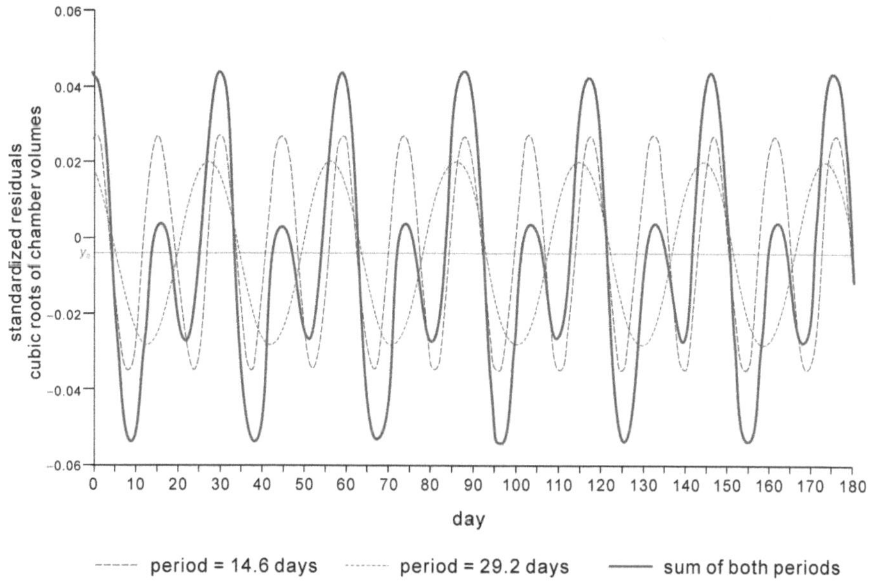

Fig. 3.20 Sinusoidal functions with the 2 larger periods obtained by REDFIT spectral analysis and sum of both functions demonstrating the intervals of two weeks with a lower peak followed by a higher peak

(6) prove again the quality of fit by Eqs. (3.6) and (3.7) and use the growth function with the better fit for calculating residuals,
(7) calculate residuals [Eq. (3.12); Fig. 3.9a] and standardized residuals [Eq. (3.13); Fig. 3.9b],
(8) calculate balanced and unbalanced outliers by fixing the 5% error limits using the mean and standard deviation of residuals (Fig. 3.16),
(9) find the mean chamber-building rate for the species under natural conditions using the Michaelis–Menten function [Eq. (3.17); Fig. 3.13],
(10) replace chamber number j by time t in the calculation of residuals using parameters of the Michaelis–Menten function [Eqs. (3.18) or (3.19); Fig. 3.15],
(11) calculate periods b of the most important sinusoidal functions [Eq. (3.20)] by power-spectral analysis using unequal time intervals (e.g., Lomb periodogram or REDFIT spectral analysis; Fig. 3.18),
(12) calculate amplitudes a and phases c of the most important sinusoidal functions by non-linear regression analyses (Fig. 3.19),
(13) interpret the results.
(14) For testing a contemporaneous onset of offspring, compare pairwise the standardized residuals of individuals based on chamber number by cross-correlation [Eqs. (3.15) and (3.16); Fig. 3.10].

References and Programs[1]

General Textbooks

Batschelet E (1971) Introduction to mathematics for life scientists. Springer, Berlin
Davis JC (2002) Statistics and data analysis in geology, 3rd edn. Wiley, New York
Hammer Ø, Harper DAT (2006) Paleontological data analysis. Blackwell, Malden
Zar JH (2010) Biostatistical analysis, 5th edn. Pearson, Upper Saddle River

Programs

Microsoft Office Excel (2010)
Hammer Ø, Harper DAT, Ryan PD (2001) PAST: paleontological statistics software package for education and data analysis. Palaeontolo Electronica 4(1):9 http://palaeo-electronica.org/2001_1/past/issue1_01.htm
PASW Statistics 19 (2010) Version 19.0.0, IBM
R version 2.15.2 (2012). The R Foundation for Statistical Computing

References

David HA (1970) Order statistics. Wiley, New York
Hohenegger J (1996) Remarks on the distribution of larger Foraminifera (Protozoa) from Belau (Western Carolines). Kagoshima Univ Res Center South Pacific, Occ papers, vol. 30:85–90
Krüger R (1994) Untersuchungen zum Entwicklungsgang rezenter Nummulitiden: *Heterostegina depressa*, *Nummulites venosus* und *Cycloclypeus carpenteri*. Dissertation, Christian Albrechts Universität Kiel
Labaj P, Topa P, Tyszka J, Alda W (2003) 2D and 3D numerical models of the growth of foraminiferal shells. In: Sloot PMA (ed) ICCS 2003, LNCS 2657. Springer, Berlin, pp 669–678
Lietz R (1996) Untersuchungen zur Individualentwicklung der Großforaminifere *Cycloclypeus carpenteri* Carpenter (1856). Dissertation, Christian Albrechts Universität Kiel
Olsson RK (1973) What is a kummerform planktonic foraminifer? J Paleontol 47:327–329
Richards FJ (1959) A flexible growth function for empirical use. J Exp Bot 10:290–300
Röttger R (1972) Analyse von Wachstumskurven von *Heterostegina depressa* (Foraminifera: Nummulitidae). Mar Biol 17:228–242
Röttger R, Spindler M (1976) Development of *Heterostegina depressa* individuals (Foraminifera, Nummulitidae) in laboratory cultures. Marit Sediments Spec Publ. 1: 81–87
Schulz M, Mudelsee M (2002) REDFIT: estimating red-noise spectra directly from unevenly spaced paleoclimatic time series. Compt Rendus Geosci 28:421–426

[1] Finally, some textbooks are mentioned that cover the field of mathematics and statistics in a broad and easily understandable sense. Afterwards, computer programs are cited that have been used in this article.

Chapter 4
Changing Investigation Perspectives: Methods and Applications of Computed Tomography on Larger Benthic Foraminifera

Antonino Briguglio, Julia Wöger, Erik Wolfgring, and Johann Hohenegger

Abstract The use of computed tomography to investigate the structure of Larger Benthic Foraminifera is presented. Several steps are required to obtain sharp and usable images with MicroCT machines. These steps are explained in detail along with select insights on larger foraminifera. This technique, which delivers three dimensional models, gives the operator the possibility to measure nearly every possible morphometric parameter on either a two- and three-dimensional basis.

Keywords Tomography • CT • Larger foraminifera • Ecology • Predation

4.1 Introduction

The informal non-taxonomic group known as Larger Benthic Foraminifera (LBF) includes symbiont-bearing species with tests commonly larger than 1 mm. Due to their very complex and beautiful tests, they have been intensively studied since the beginning of the nineteenth century. The studies on their morphologies in oriented sections (i.e., along their equatorial or axial planes) led to a broad spectrum of taxonomic results, important for biostratigraphy to paleoecology in shallow water tropical to warm temperate marine paleo-environments.

A. Briguglio (✉)
Natural History Museum Vienna, Burgring 7, 1010 Vienna, Austria

Department of Paleontology, University of Vienna, Althanstrasse 14, 1090 Vienna, Austria
e-mail: antonino.briguglio@univie.ac.at

J. Wöger • J. Hohenegger
Department of Paleontology, University of Vienna, Althanstrasse 14, 1090 Vienna, Austria

E. Wolfgring
Department of Geodynamics and Sedimentology, University of Vienna,
Althanstrasse 14, 1090 Vienna, Austria

H. Kitazato and J.M. Bernhard (eds.), *Approaches to Study Living Foraminifera: Collection, Maintenance and Experimentation*, Environmental Science and Engineering,
DOI 10.1007/978-4-431-54388-6_4,

In fact, in LBF the measurement of linear parameters on equatorial section often leads to precise taxonomic identifications. The extraordinary evolution through geological time of morphometric characters allows a given taxa to be significant markers for very short time intervals. Therefore, certain LBF are considered as important stratigraphic index fossils during specific time intervals on the geological time scale.

Axial sections are also very important to distinguish foraminifera. Sometimes, as in the case of the fusulinids or orbitolinids, they are characteristic for species identification. However, axial sections are widely used to get information about environment and ecology. Surface/volume ratios as well as diameter/thickness ratios are routinely used parameters to infer water depth, the hydrodynamic scenario and depositional conditions (Beavington-Penney et al. 2005). Unfortunately, due to the sectioning process, applied to obtain oriented sections, parts of the test are lost and the necessary combination of information from the equatorial and axial view on the same specimen is impossible. As in the case of thin sections from bulk samples such as limestones (abundantly used for facies analyses), the possibility to center an oriented section is very low, time consuming and the rest of the test is always destroyed.

Recently, a new technique allowed observation of objects in all possible views, to measure them and, most importantly, not destroy them: the computed tomography (CT). Computed tomography is an X-ray based technique that allows visualization of internal structures of objects permitting radiation to pass through. Depending on the working stations used, the resolution of the image and the penetration of radiation can be very different, thus delivering different results. The possibility to use CT working stations for studying tests of LBF is a great advantage as the visualization of such complex tests has been always hampered by different two-dimensional approaches explained above. Several works have been already done proving the potential of CT scans on foraminifera (Speijer et al. 2008) and illustrating some basic approaches (Görög et al. 2012). Most recently, CT scans of LBF have been used to make three-dimensional models where several two and three dimensional parameters may be measured to get information on ontogeny (Briguglio et al. 2011) and on some paleobiological adaptations (Briguglio et al. 2013). Besides measuring morphometric parameters, several evidences and questions arise by the simple three-dimensional visualization of complete tests. Injuries, recoveries, multilocular apparatii, adult twins are consistently present in several populations we have scanned to date. In this paper, attention is focused mainly on the scanning methodology specifically for LBF tests, but special attention is also given to some aspects that provide rich research direction for the near future.

4.2 The Scanning Process

4.2.1 The Scanner

CT scanners are quite simple instruments composed mainly of an X-ray source represented by a (micro) focused X-ray tube, a charged-coupled device (CCD)

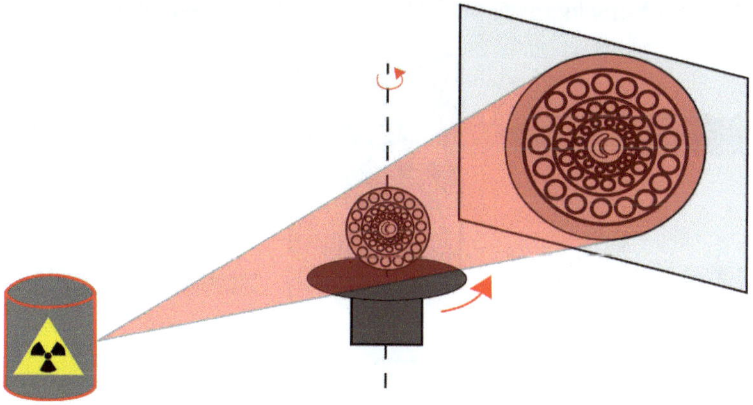

Fig. 4.1 Simplified scanning mechanism composed by an X-ray source (*left*), a rotating object holder (*center*) and a CCD detector (*right*)

camera such as the X-ray detector, and a sample tray (Fig. 4.1). During the scanning process, several one- or two-dimensional radiographs are recorded for different positions during a stepwise rotation around a central axis of either the sample or the system source-detector. Commonly, in scanners built for academic purposes, the sample rotates and the system source-detector remains fixed, while in scanners built for medical purposes (where the "samples" are human bodies), the source-detector system rotates around the sample.

Excluding those for medical purposes, there is a large variety of scanners available for scientific purposes and they vary according to scan size, imaging intensity and image resolution. While larger detectors allow the scanning of larger objects within one single procedure, the intensity of X-ray radiation discriminates the kind of material suitable for CT investigation. In fact, penetration intensity is inversely proportional to material density and specimen thickness (Fig. 4.2). Thus, high intensity X-ray sources are able to image thicker object(s) with relatively higher density. Recent LBF do not require any special high intensity X-ray source to be sufficiently imaged. Fossil LBF forms do not require any high intensity X-ray source either.

Concerning resolution, CTs are classified within three main groups: CTs, microCTs and nanoCTs. Commonly, nanoCTs have a detectability below 500 nm (maximum range 150–1,000 nm) but the sample size commonly not exceed 11 mm in diameter; microCTs have a detectability down to 5 μm (maximum range 3–300 μm), and can host larger specimens (so far, not larger than 10 × 10 × 10 cm), normal CTs, mainly used for medical purposes may treat human-body size despite resolution that is normally in 1–2.5 mm range. Regarding image resolution, most of the CT scanners have several built-in CCD cameras characterized by different resolution.

Because computers require considerable time to record high-resolution images, scans taken by high-resolution cameras are normally time consuming (thus, often more expensive) and require large storage capacity in bytes. Depending on the tasks and on the scientific goals of the scan, the decision to use a different camera might

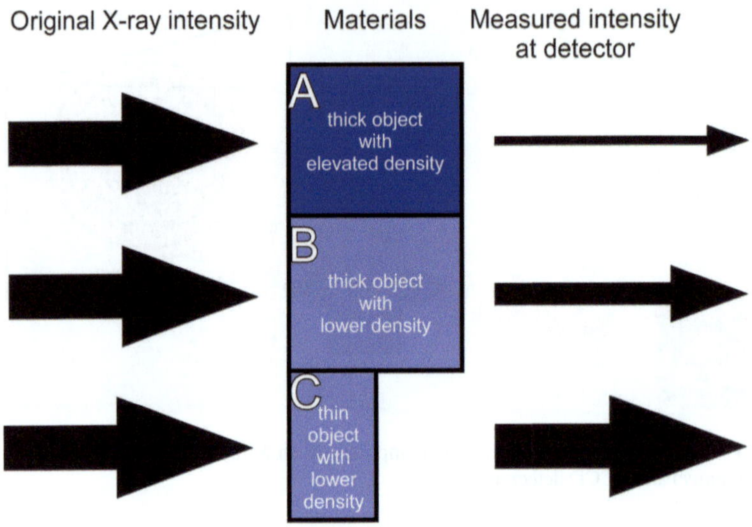

Fig. 4.2 Differences in X-ray intensity before and after penetration. Depending on thickness and density of the material to scan, X-ray intensity might not be detected by the CCD sensor

be pivotal to get the desired resolution, especially for larger objects or to make trial scans to check for density contrast. For LBF, the use of high resolution CTs (micro or nano) is commonly mandatory since the resolution needed to visualize chambers or parts of chambers are always <0.5 mm. Depending on the scanner, especially for nanoCTs, the use of low resolution cameras might still allow fast, good-quality scans.

4.2.2 *Sample Preparation*

One of the most common advantages of the CT technique is that it does not require any special sample preparation or chemical fixation. Nevertheless, proper methodology and some tricks do help for good visualization and clear imaging. The most important parameter to consider for successful scans is the density of the specimen of interest. CT scanners can only produce usable results if the features that must be represented are marked by sufficiently great density differences. For scans of LBF, the density of the test should be markedly different from the density of the surrounding sediments and from the material filling the cavities. Therefore, calcareous hyaline tests preserved in limestones are very poorly suited for CT investigations because the density of the calcium carbonate of the LBF test is very similar to the density of carbonate sediments. Similarly, hyaline tests of LBF with chambers filled by calcite will not be revealed by microCT scanning. Fossil samples where the filling is either glauconite or pyrite (both common) will give adequate results. Self-evident, isolated empty tests will give best results. For microgranular tests (e.g., fuslinids, orbitolinids), depending on the filling material, the scan may still be adequate but this strongly depends on the density of the foraminiferal test.

Securing the sample onto the sample holder is a delicate procedure for which several circumstances should be considered. The samples should be placed in standing position and not in lying position. The standing position allows the X-ray to cross the sample along shorter distances and therefore not to be deviated, refracted or attenuated by too much material. In fact, rotation of a LBF specimen during the scanning procedure will reveal its equatorial view most of the time instead of its axial view, which is mostly thicker and more complicated to reconstruct (see Sect. 4.2.4).

Only few ways to fix the specimen on the sample holder have proven useful throughout the scanning process. The use of plasticine as the basis to adhere part of the sample to let it stand should be avoided since such material is not transparent to X-ray; thus a portion of the sample will not be recorded by the detector because plasticine is imaged as a black surface. Very small plastic pipettes work very well to solve this problem. The lower part can still be attached to the specimen holder by plasticine but the lower part of the pipette can be easily filled with some fragments of paper tissue which is transparent to X-rays and where the sample can be positioned to avoid damage or breakage. Folded cardboard is excellent to help specimens stand and prevent them from falling down during the scanning process. To avoid specimens moving during the scan, some supplemental plasticine on the top of the plastic pipette might be a good solution to push down the specimens. The use of some additional paper tissue between cardboard and plasticine insures that the specimen is not on the same level of the plasticine and thus the specimen is continuously imaged during the rotation process (Fig. 4.3).

4.2.3 Scanning Process and Acquisition of Data

Either fully automated or manually controlled, the scanning process represents the most important step to obtain focused images of the surface and the interior of an object.

Although there are different companies who produce CT systems and each of them has different software; the parameters to be controlled by the operator to get good and focused results are very similar in all systems and can be listed as: X-ray intensity (kV and mA), rotation step parameter, averaging parameter, magnification (i.e., resolution) and filter use.

The best X-ray intensity has to be chosen for each object to scan as it can be different according to sample size (mainly thickness) and density. Choosing the right X-ray intensity for a sample to image is a semi-automated procedure where the operator can visualize in real time the amount of radiation (i.e., penetration) measured by the detector. To obtain high quality results, the penetration of the X-ray beam through the sample should be comprised between 30 % and 80 % of the penetration of the same beam trough the air. Profile lines calculation is a standard tool available on all scanning software. Due to the rotation, which operates during the scanning process, many samples may reveal a greater thickness to be penetrated at certain rotation angles (e.g., equatorial view vs. axial view of a LBF); it is important to check the penetration rate at such view angles before starting the scanning process. For recent LBF, an X-ray intensity of 100 kV and 80 μA often gives

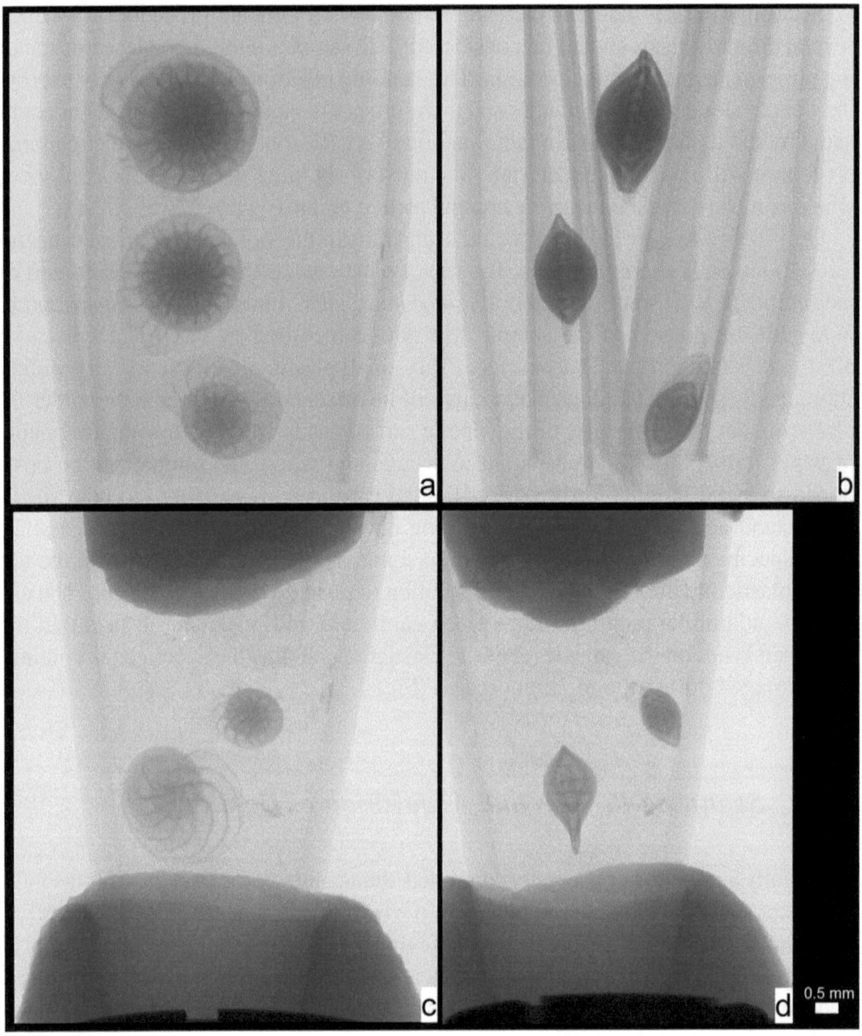

Fig. 4.3 X-ray images of foraminifera prepared to be CT scanned. Frontal view of three recent foraminifera prepared to not overlap during scan procedure (**a**). Cardboard maintains foraminifera placement during the scan procedure (**b**). Same procedure without cardboard but with paper tissue (**c**, **d**). In all cases, plasticine (*the darker shadow above* and *below*) has been used to secure the sample. All samples have been located in a plastic pipette

excellent results. Fossilized specimens might require higher intensity (e.g., up to 150 kV). In general, higher intensity (i.e., higher penetration) should be avoided as image oversaturation will compromise the quality of the results.

The "rotation step" parameter defines the rotation degree at which the camera will record one image or several images. The smaller this number, the higher the scan time and data set size will be (Fig. 4.4). Furthermore, a lower rotation-step

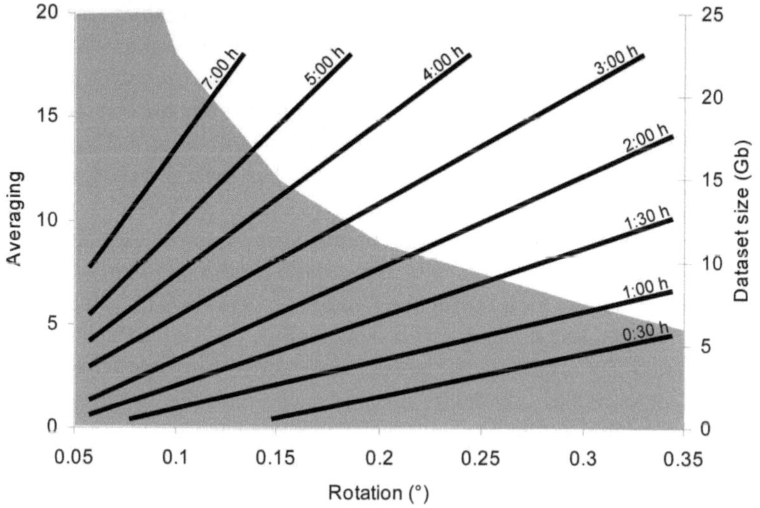

Fig. 4.4 Scanning parameters. Depending on the rotation and averaging, scan duration (*straight lines*) varies strongly. The averaging does not affect the dataset size (*grey area*), which is only influenced by the rotation step parameter

value will increase the visualization of details on the three dimensional model. For LBF, values between 0.10° and 0.20° will give high quality results. Depending on the general resolution of the scan—which will be discussed later—rotation step values lower than 0.10° might not improve the quality of the results but will only increase the scan time which might be directly correlated to both machine usage costs and radiation source life time.

The averaging parameter defines the number of frames to be averaged for the rendering of a single two-dimensional radiography. An increase of the averaging parameter will directly increase the scanning time but not affect the dataset size (Fig. 4.4). More focused images can be observed by increased averaging, especially in external details and moreover in samples where several materials (i.e., several densities) are involved.

High averaging values might in fact differentiate materials possessing slightly different densities. For the scan of recent LBF tests, the materials to be imaged are calcareous or agglutinated tests and air filling the empty spaces (i.e., the chambers) only. For this kind of sample, an averaging value computed between 2 and 5 gives well-focused images and reduces acquisition time drastically. For fossil tests possessing calcareous walls and calcite fillings, very high averaging values (above 50) might poorly differentiate some structures. Commonly, the results will be mostly unusable and the scan time will last for days. Fossilized specimens with some pyrite fillings or where small structures (canal system in LBF) are recrystallized might still be imaged with higher averaging resulting in focused three dimensional models. Scans with high averaging values are often used by scientists studying soft-tissue preservation in both recent and fossil forms to better differentiate organs and organic tissues.

The magnification of a scan depends mainly on the CCD camera resolution and on the position of the sample along the line defined by the system "X-ray source-object-detector". Modern scanners are equipped with different CCD cameras characterized by different resolutions to deliver scans at different qualities and different speeds. A low resolution camera is very useful for trial shots to test the X-ray penetration potential of an object. Once the camera has been selected, resolution is directly related to the distance of the sample to the X-ray source. The resolution increases by moving the object closer to the X-ray source. Particularly, larger objects (agamonts in LBF) might not be imaged too close to the source because they could exceed the field of view during the rotation process. In CT scanners, the resolution is given in pixels depending on the detector type, but on the three-dimensional models obtained the word "voxel" is used instead of pixel. The voxel is a three-dimensional pixel, it has commonly a cubic shape and, therefore, the pixel size represents the length, the height and the thickness of the voxel. To calibrate the three-dimensional dataset obtained by the scanner, the voxel size is the most important parameter.

Before starting the scan process, some scanners are provided with filters that can enhance the quality of the results. Filters are commonly metal plates covering the X-ray source and are called "beam hardening filters" as they influence radiation. The main duty of metallic filters is either to reduce specific parts of the X-ray radiation or to clean it by cutting some frequency radiation peaks. Copper filters as well as aluminum filters are well known filters for CT scanners (Ay et al. 2012). Thick aluminum filters (e.g., 1.0 mm thickness) cut off the low-frequency spectrum allowing only its highest part to image the object, thus reducing large parts of radiation that will not penetrate the specimen. The use of filters is very important since the photons produced by the X-ray tube are radiated in a number of energies, some very low and others very high. For scanning purposes, it is important that all photons reach the CCD detector at different intensities to get two-dimensional radiographic information. Since low-energy radiations (i.e., photons) are more easily absorbed by low-density material (e.g., soft tissues in recent or living organisms) and do not reach the detector, it is always preferable to reduce the output radiation to the most energetic one. Since aluminum allows penetration of high energy X-rays only, the placement of aluminum plates in front of the X-ray tube will reduce part of the radiation beam. For LBF, the use of such filters is highly recommended because foraminiferal tests will only be revealed by the detector if imaged by high energy X-rays. The low-energy radiation will only increase the general noise.

4.2.4 Reconstruction Process

During the scanning process, a large amount of data is recorded. Some thousands of images, commonly ".tiff" files, have now to be assembled to produce three dimensional models: this is done through a so called "reconstruction process". This is an algorithmic-based procedure, which is commonly made by a computer cluster equipped with larger storage and memory systems. Extreme computational power is

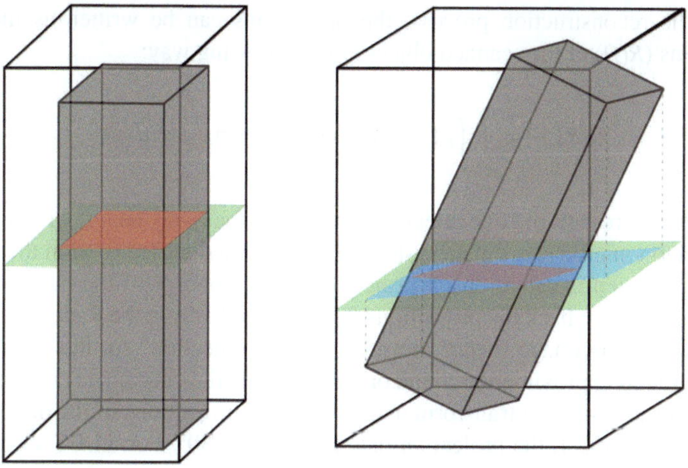

Fig. 4.5 Reconstruction process. The vertical object in the center is reconstructed by the sum of the horizontal slices in vertical succession. The large horizontal plane represents the ROI. When the sample has been scanned in vertical position, the ROI can be reduced (almost) to the size of the small horizontal slices. A 25° rotated position of the object increases the sufficient ROI to a much larger area, since all object's projections must be included into the ROI. The larger is the ROI, the longer will be reconstruction time and the larger the resultant dataset

needed because the reconstruction process must collect information stored by the CCD detector and reveal slices normal to the X-ray imaging plane (the horizontal planes in Fig. 4.5).

The data manipulation possibilities for the operator to get a successfully reconstructed dataset are few. A large part of computational options is mainly processed by the computer. However, the few adjustments that the operator might insert in the process are easily understood if some basic mathematics about the reconstruction process itself is explained. There are different methods the computers use to run reconstructions and, depending on the algorithm, the whole process might be more or less rapid.

During the last decade, the most frequently used algorithm was written by applying the "Projection Slice Theorem" also called the "Central Slice Theorem" first reported by Bracewell (1990) and based on the Radon Transform (Radon 1986). This theorem considers that if an infinite number of one-dimensional projections of an object are available at in infinite number of angles, the reconstruction of the original object will be perfect. Radon Transform allows this by the reconstruction of geometric shapes in two dimensions calculating integrals of functions over straight lines. The more of those lines, the better the reconstruction accuracy will be. Geometrically, the projection of an object at a certain angle over a two dimensional surface can be calculated as a set of line integrals. In computed tomography, the integrals are constituted by the intensity of the X-ray beam measured by the CCD camera after the beam itself has been attenuated by the object scanned along a single line on a surface (i.e., the detector plane).

For the reconstruction process, the projections can be written as the Radon transforms (*R(f)*) of the scanned object in the following way:

$$R(f) = \int_{-\infty}^{\infty}\int_{-\infty}^{\infty} f(x,y)\, f(x\cos\theta + y\sin\theta - s)\, dx dy \tag{4.1}$$

where *R(f)* represents the line integral along a line (the X-ray beam) within an established trigonometric coordinate system (θ, s) and at a distance $|s|$ from the origin.

The next step is called "back projection". The concept of the back projection algorithm is to run back the projections (i.e., the results from the Radon transformation) through the image (hence the name "back projection" for this operation) to obtain a rough approximation to the original in a spatial coordinate system.

In order to invert the transform and to get back projections of the data recorded at the CCD detector, the projection data must span 180°. For all CT scanners, the standard rotation is, therefore, 180°.

The back projection function *B(g)* can be written as:

$$B(g) = \int_{0}^{\pi} g(x\cos\theta + y\sin\theta)\, d\theta. \tag{4.2}$$

The back projection function represents the accumulation of the projection passing through the point (x, y) for each angle θ for a complete rotation of 180° (0 to π). Due to this operation, a blurred image is revealed and can be affected by star-like or ring-like artifacts (depending on the rotation step parameter value; star-like if it is very high, ring-like if it is very low) occurring at 180° from the reconstructed image. A two-dimensional ramp filter, which cuts out low frequencies and lets high frequencies pass, is well suited to clean such artifacts and is included in all reconstruction algorithms. Modern scanners allow 360° scan procedures to reduce the artefact rate, although this full-circle approach doubles both scan time and dataset size.

As reported in Fig. 4.4 the size of the dataset after the scanning process, depending on the scan parameters, might be very large. The reconstruction process reduces it, but several tricks help in reducing the size of the output data. Almost all reconstruction programs allow the operator to modify arbitrarily the volume to reconstruct by the delimitation of a "region of interest" (ROI). The smaller the volume is to reconstruct, the faster the reconstruction process and the smaller the output dataset will be, thus, easier to be manipulated on a computer. For LBF, which commonly possess symmetric shapes (e.g., flat-, spindle-, lentil-shaped) the ROI might be strongly reduced. In fact, the highest reduction of the ROI is achieved if the specimen has been scanned in vertical standing position. If the specimen has not been positioned vertically, the ROI must be enlarged to be sure its lateral projection fits in it (Fig. 4.5).

Before starting the reconstruction process it might be important to reduce the grey scale values to those relevant for the manipulation of the results. By enlarging the grey-scale spectrum, much more detail becomes visible and, thus, can be considered during the reconstruction process. However, some of these details might not be important or not be part of the scanned specimens. For example, if the scan has been run by fixing the specimen in paper tissue and/or cardboard, these will be revealed by a larger grey scale spectrum. Therefore, it is important to carefully reduce the grey scale values to clean external parts not belonging to the scanned object but without erasing parts of the object itself.

4.2.5 Slicing Process

Depending on the calibration of the reconstruction algorithm and on the size of the ROI, the reconstruction process might take longer time and produce larger datasets. In our example, considering the scan was run on a vertical standing specimen and the ROI has been limited as explained, the output files produced by the reconstruction process are now a series of rectangular slices parallel to the axial section of the specimen. These files are now ready to be used in all visualization programs and can be rendered to plot a three-dimensional model. However, in some cases, the reconstruction files may remain very large requiring sophisticated computational systems. Further reduction of the dataset without affecting the quality of the results is possible and always suggested. Included in all CT software packages, there are some applications for three-dimensional data view on a two-dimensional basis. Those programs allow the rotation of the two dimensional projections of the specimen on three different axes. Furthermore, they allow the recreation of the dataset by slices parallel to each one of the three planes displayed. These functions are most important to reduce the size of the dataset. In fact, the best way to reduce the size of the dataset is to delete those parts of volumes that not include the specimen. An accurate rotation of the specimen in vertical position permits to extract a volume of interest (VOI) which only includes the specimens and not the surrounding volumes. Once the VOI has been adjusted, the specimen can be resliced along one of the three views. The most convenient view to slice it is the one containing the fewest number of slices. For nummulitids, the equatorial view suits this request; for alveolinids or fusulinids it will be the axial one. Such an easy procedure only cuts volume data outside the scanned object and does not affect the quality of the scan, resulting in a much lighter dataset, normally never exceeding 1 GB in size.

4.2.6 3-D Rendering

At the end of these procedures, the dataset is ready to be used. Dedicated computer programs allow three-dimensional renderings of volumes, surfaces, interiors and much more. Some tools allow selection and exportation of specific parts of a three-dimensional model: this procedure is called segmentation. Segmentation is based on grey scale values. Semi-automatic tools search for connected intervals with equal grey scale values and export the volume they represent. In LBF this is a great tool to extract chambers, chamberlets, canal systems and microboring evidence. Some extra tools allow the operator to cut a large connected region to separate two or more three-dimensional objects. In LBF, all chambers are connected to each other by foramina and/or stolons. Therefore, if the extraction of the lumina is requested, a simple click on the first chamber will automatically extract all chambers possessing the same gray scale value, which if they are empty it is "zero" since they are completely black. If the differentiation of each chamber is requested, then the operator should manually disconnect all chambers slice by slice.

Depending on the resolution of the dataset and on the gray-scale interval chosen to extract the chambers, those connections might be not always visible and the segmentation process is a "one click procedure" per chamber. Since the data set can be calibrated by using voxel size, always given after each scan, the whole three-dimensional model as well as part of it (i.e., segmented volumes) might be measured in all possible directions. Distances, angles, surfaces, volumes, center of gravity and surface roughness are just examples of the diversity of measurements that dedicated -D graphic computer programs might run on such datasets. Morphometry in LBF is a fundamental basis for taxonomy and ecology. Some results of microCT measurements on LBF are reported in Briguglio et al. (2011, 2013), Hohenegger and Briguglio (2012) and in this book (Hohenegger and Briguglio 2014). Since the dataset may be visualized by all scientists, the distribution of these can be really important in case of holotype specimens. Recently, a new species of LBF, a rotaliid from the Oligocene, has been described and published; for the first time in science, a complete stack of slices parallel to the equatorial and on the axial plane of the holotype have been updated as repository data (Benedetti and Briguglio 2012) and all scientists may look at the data and create cuts and slices along whatever direction and observe details not reported in the species description if their research is focused on other information.

4.3 Latest Results

As a practical aspect of this new technique, some latest results are presented to show the variety of applications that CT enables. The capability to observe (and measure) the totality of the internal and external morphology of a LBF is unique and so far has never been explored. Thin-section methodology always allowed the visualization of only one "slice" of the entire specimen. The first example reported is based on 14 specimens of *Cycloclypeus carpenteri* collected at 50-m water depth offshore from the island Ishigaki (Japan), some of these are illustrated in Fig. 4.6. They were all living specimens at the time of collection due to their intensive brownish color given by their symbionts. Externally, no major morphological variations were visible except for 3 specimens that were characterized by a light saddle

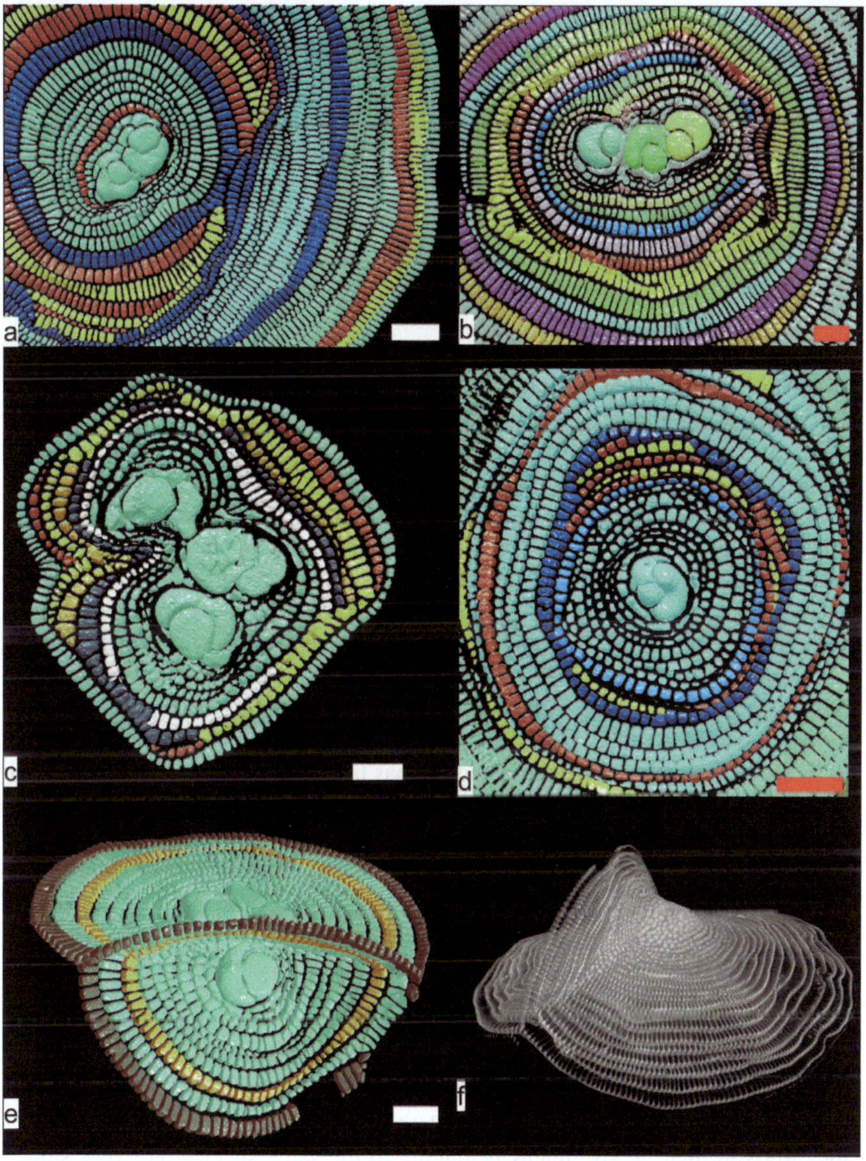

Fig. 4.6 Selection of investigated *C. carpenteri* specimens. Specimen A10, multiembryonic apparatus with two embryos, six broken chambers are recovered (**a**); Specimen A2, multiembryonic apparatus with three embryos, several breakages are evident in the first chambers (**b**). Specimen A5, multiembryonic apparatus with three embryos, the first five chambers are all broken (**c**). Specimen A7, seven chambers are broken and are sutured by chamber 15 (**d**). Specimen A6, multiembryonic apparatus with three embryos and two connected equatorial layers, the first seven chambers of both planes are not broken (**e**). Specimens A3, external view of a specimen with multiembryonic apparatus with two embryos and two connected equatorial layers, major breakages are present on the vertical layer (**f**) (see Fig. 4.7). Scale bar: 0.5 mm and the longest diameter of specimen A3 is 12 mm

Table 4.1 Observation data on the *C. carpenteri* dataset

Specimen's name	A1	A2	A3	A4	A5	A6	A7	A8	A9	A10	A11	A12	A13	A14
Chamber number	30	26	35	31	19	29	36	24	28	30	28	18	35	30
Broken chambers %	13	27	43	32	68	52	50	22	57	57	39	39	49	21
Max. recovery	2	6	9	3	3	5	6	2	7	7	3	4	2	3
Proloculi	1	3	2	1	3	3	1	1	1	2	1	1	1	2
Equatorial planes	1	1	2	1	1	2	1	1	1	1	1	1	1	1

See text for explanation

shaped test (specimen A14), or by a second equatorial layer (specimens A3 and A6). After visualization by micro CT, the morphological interpretation changes conspicuously (Table 4.1). The explanation for all morphological terms hereafter used, can be found in the "Illustrated glossary of terms used in foraminiferal research" (Hottinger 2006).

Almost half of the investigated specimens (i.e., 42 %) possessed a multiembryonic apparatus with two or three proloculi and deuteroloculi and consequently their relative nepionic chambers (Fig. 4.6a–c, e, f). In only two cases, multiple equatorial layers have been observed, always connected to multiembryonic apparatii (Fig. 4.6e, f). It has been observed that such multiple equatorial layers are growing synchronously. The same chamber wall can be traced continuously on both layers (Fig. 4.6e, f). Most interesting is the evidence that each specimen has experienced chamber damage during its life (Fig. 4.6a–d). In fact, in all investigated specimens, an average of 40 % of their chambers exhibit damage. It has been also observed that each specimen was able to recover after each injury, evidenced by addition of new test chambers. The most impressive recovery has been observed in a secondary equatorial layer in sample A3 (Fig. 4.6f) where one complete chamber repairs nine broken chambers (Fig. 4.7).

A second analysis shows the possibility to study growth allometry and functional morphology by the measurements of volume and surface of each chamber during growth (Fig. 4.8). The data reported have been collected after segmentation of 505 chambers (proloculus and deuteroloculus have been always excluded) belonging to 11 specimens of *Palaeonummulites venosus* and of 285 chambers of ten specimens of *C. carpenteri*. Such plots may be very helpful to better quantify test morphology of single specimens or of the entire population in relation to their environment. It is not surprising that all specimens belonging to the same species plot very similar to each other, but is very interesting how these different geometries (annular for *C. carpenteri*, and spiral for *P. venosus*) reflect the different environments where the two species live. In fact, *C. carpenteri* can be found alive from 50 to 90 m water depth, while *P. venosus* is particularly abundant at 30–40 m (among others Hohenegger 2004). The light intensity conditions can be considered one factor leading to such different volume to surface area ratios.

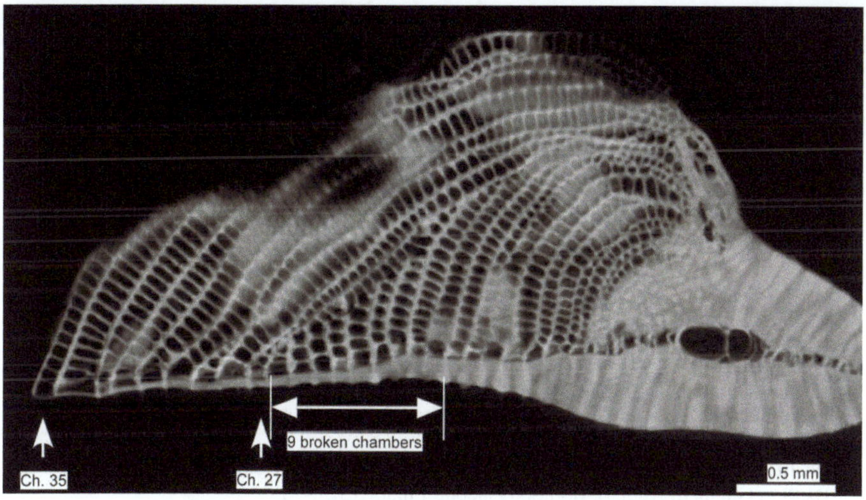

Fig. 4.7 *C. carpenteri* specimen A3. Largest recovery evidence recorded. Chamber 27 is built on nine broken chambers

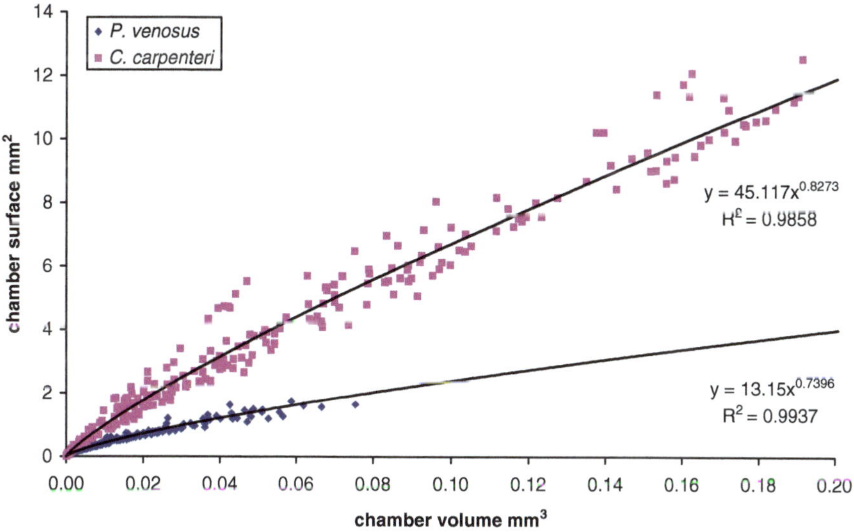

Fig. 4.8 Chamber volume versus chamber surface for 11 specimens of *C. carpenteri* (285 chambers) and for of 11 specimens of *P. venosus* (505 chambers). Power functions have been calculated on each dataset

Acknowledgements This work was developed within the project P 23459-B17 "Functional Test Morphology of Larger Benthic Foraminifera" of the Austrian Science Fund. We thank the Department of Paleontology for making the use of MicroCT possible and for providing a dedicated working station to analyze the datasets. We thank Joan Bernhard for her constructive comments and suggestions.

References

Ay MR, Mehranian A, Maleki A, Ghadiri H, Ghafarian P, Zaidi H (2012) Experimental assessment of the influence of beam hardening filters on image quality and patient dose in volumetric 64-slice X-ray CT scanners. Phys Med 28:1–12

Beavington-Penney SJ, Wright PV, Racey A (2005) Sediments production and dispersal on foramifera-dominated early tertiary ramps: the Eocene El Garia Fm, Tunisia. Sedimentology 52:537–569

Benedetti A, Briguglio A (2012) *Risananeiza crassaparies* n. sp. from the Late Chattian of Porto Badisco (southern Apulia). Boll Soc Paleontol Ital 51(3):1–10

Bracewell RN (1990) Numerical transformation. Science 248:697–704

Briguglio A, Metscher B, Hohenegger J (2011) Growth rate biometric quantification by X-ray microtomography on larger benthic foraminifera: three-dimensional measurements push nummulitids into the fourth dimension. Turk J Earth Sci 20:683–699

Briguglio A, Hohenegger J, Less G (2013) Paleobiological applications of three-dimensional biometry on larger Benthic Foraminifera: a new route of discoveries. J Foramin Res 43(1): 67–82

Görög Á, Szinger B, Tóth E, Viszkok J (2012) Methodology of the micro-computer tomography on Foraminifera. Pal Elect 15(1):3T

Hohenegger J (2004) Depth coenoclines and environmental considerations of western pacific larger Foraminifera. J Foramin Res 34(1):9–33

Hohenegger J, Briguglio A (2012) Axially oriented sections of nummulitids: a tool to interpret larger benthic Foraminiferal deposits. J Foramin Res 42(2):145–153

Hohenegger J, Briguglio A (2014) Methods for estimating growth pattern and lifetime of foraminifera based on chamber volumes. In: Kitazato H, Bernhard JM (eds) Approaches to study living foraminifera: collection, maintenance and experimentation. Springer, Tokyo. doi:10.1007/978-4-431-54388-6

Hottinger L (2006) Illustrated glossary of terms used in foraminiferal research. Notebooks Geol, Mem 2006/2. http://paleopolis.rediris.es/cg/uk_index.html_MO2 pdf 1–4

Radon J (1986) On the determination of functions from their integral values along certain manifolds. Trans Med Imag MI-5(4):170–176

Speijer RP, Van Loo D, Masschaele B et al (2008) Quantifying foraminiferal growth with high-resolution X-ray computed tomography: new opportunities in foraminiferal ontogeny, phylogeny, and paleoceanography applications. Geosphere 4:760–763

Chapter 5
Protein Analysis in Large Benthic Foraminifera

Steve S. Doo, Anderson B. Mayfield, Hong D. Nguyen, and Hung-Kai Chen

Abstract Large benthic foraminifera (LBFs) have long been used as environmental recorders of ocean chemistry. Although the importance of foraminifera in paleo-reconstructions of ancient oceans and as sediment producers is well documented, the biology of tropical symbiont-bearing foraminifera has only recently gained increased attention. Tropical symbiont-bearing LBFs represent a unique and important subset of LBFs in that they are vital to coral-reef ecosystems and host a wide suite of algal symbionts (e.g., dinoflagellates, diatoms, red algae, green algae and cyanobacteria). Previous studies on both host and symbiont physiology have been performed in order to gauge the foraminiferal response to a variety of stressors, including elevated temperature and nutrient levels, as well as acidification. Recently, protocols have been developed for protein analysis in LBFs that will allow for expression analyses of target proteins from both members of the holobiont. In this chapter, we detail a protein expression protocol for one-dimensional sodium dodecyl sulfate poly-acrylamide gel electrophoresis (1-D SDS-PAGE) and consequent western blotting for determination of protein expression in the foraminiferal holobiont.

S.S. Doo (✉)
School of Biological Sciences, University of Sydney, Sydney, NSW, Australia

National Museum of Marine Biology and Aquarium, 2 Houwan Road, Checheng, Pingtung County (R.O.C) 944, Taiwan
e-mail: steve.doo@sydney.edu.au

A.B. Mayfield
National Museum of Marine Biology and Aquarium, 2 Houwan Road, Checheng, Pingtung County (R.O.C) 944, Taiwan

Living Oceans Foundation, Landover, MD, USA

H.D. Nguyen
School of Medical Sciences, University of Sydney, Sydney, NSW, Australia

H.-K. Chen
National Museum of Marine Biology and Aquarium, 2 Houwan Road, Checheng, Pingtung County (R.O.C) 944, Taiwan

H. Kitazato and J.M. Bernhard (eds.), *Approaches to Study Living Foraminifera: Collection, Maintenance and Experimentation*, Environmental Science and Engineering, DOI 10.1007/978-4-431-54388-6_5,

This technique has the potential to target proteins that are specific to either host or symbiont compartments, a breakthrough that may ultimately allow for an increased understanding of the molecular-scale regulation of the symbiosis that is vital to these globally important calcifiers.

Keywords Protein expression • SDS-PAGE • Western blotting • Large benthic foraminifera

5.1 Introduction

Large benthic foraminifera (LBFs) are important producers of marine sediments on tropical coral reefs, contributing to greater than 80 % of the calcium carbonate production in certain areas (Langer 2008). Accumulated tests of post-mortem foraminifera are also widely used in microshell geochemistry and shell taphonomy for paleo-reconstructions of past ocean conditions. Furthermore, the influence of foraminifera tests and their role in buffering of diel pH changes on tropical coral reefs are just beginning to be understood (Yamamoto et al. 2012). Although the importance of foraminifera post-mortem tests is well documented, the biology of these protists has been vastly understudied, particularly with respect to their symbiotic associations with a wide suite of marine algae (Lee 2006). In this mutualistic symbiosis, the protozoan host relies heavily on the algal symbionts to meet its metabolic demands (Fujita and Fujimura 2008). As such, the physiological performance of the symbionts can greatly affect the overall health of the host, as well as the holobiont as a whole. In the context of changing marine climates, understanding how each member of this symbiosis responds to forecasted ocean conditions will be key to predicting how foraminifera will fare (Lee 2006; Hikami et al. 2011).

The physiological response of foraminifera to a wide suite of various stressors, including temperature, acidification and eutrophication, have been explored in studies based on parameters that assessed the performance of the holobiont as a whole. Measurements such as respiration rate, intra-organismal pH, growth, and surface area are difficult to partition into discrete host and symbiont compartments and are, therefore, typically reported for the entire holobiont. That being said, some commonly measured parameters, such as the maximum quantum yield of photosystem II (F_V/F_M), chlorophyll a (*chl-a*) concentrations, and symbiont density are direct measures of the algal contribution to the holobiont (e.g., Schmidt et al. 2011; Sinutok et al. 2011).

Techniques involving the molecular biology of foraminifera have been notoriously difficult, and in many cases, universal eukaryotic primers for polymerase chain reaction (PCR) are not compatible with these foraminifera. Recently, breakthroughs in LBF molecular biology, particularly compartment-specific protein expression, have provided new insights into the physiology of the protozoan hosts and algal symbionts through investigation of proteins unique to each member of the holobiont (Doo et al. 2012; Heinz et al. 2012). In this chapter, we present a protocol for low–medium throughput analysis developed specifically for foraminifera that potentially allows for targeted analyses of either host foraminifer or symbiotic algae protein expression. This protocol is adapted from traditional methods of protein extraction/isolation, gel electrophoresis and immuno (western)-blotting. We also describe potential future applications of this technique that may aid in molecular-level analyses of this marine symbiosis.

5.2 Collection, Preservation and Protein Extraction

Equipment and reagents
Microcentrifuge tubes
96-well microplates
Spectrophotometric microplate reader (562 nm)
Pierce® BCA Protein Assay kit
Microcentrifuge (4 °C)
Liquid nitrogen
Radioimmunoprecipitation assay (RIPA) buffer with protease inhibitor (see Sect. 5.2.2.)

Note: The majority of the chemicals utilized in these protocols (e.g., acrylamide) are hazardous and require the use of proper protective gear (lab coats, gloves and safety goggles). Furthermore, homogenization and mixing of chemicals should be conducted in a fume hood. Waste products should be disposed of in accordance with local laws to minimize impacts on the environment.

5.2.1 *LBF Culturing and Sample Size Considerations*

Generation of LBF cultures provides the samples needed to perform manipulative experiments under controlled laboratory conditions. Standard conditions of LBF culturing should be adhered to, with recommended photosynthetically active radiation (PAR) levels between 100–300 μmol quanta m^{-2} s^{-1} based on previous experiments of photosaturation in symbiotic planktonic foraminifera (Rink et al. 1998). When possible, LBFs should be reared in an open system, as flow has been shown to affect foraminiferal growth rate (Glas et al. 2012).

The extraction of protein at sufficient concentrations for molecular analyses is crucial for successful analyses. In our experience, 10–20 smaller calcarinid foraminifera (e.g., *Baculogypsina sphaerulata*) with healthy symbionts will yield ~15 μg protein (Doo et al. 2012), a suitable quantity for SDS-PAGE and western blotting (Table 5.1). Other studies have used larger numbers of LBFs, ~100–150, and were able to purify >50 μg total protein (Heinz et al. 2012; Table 5.1).

Table 5.1 Calculated total protein per individual foraminifer indicating a large range of values between and even among species

Species	Location of samples	# Total measurements (# Total specimens)	Total protein per individual (μg)	References
Ammonia tepida	Bay of Aiguillon, France	4 (7,070)	0.066±0.012	Heinz et al. (2012)
Massilina secans	Yeu Island, France	3 (315)	0.231±0.34	Heinz et al. (2012)
Ammonia beccarii	Yeu Island, France	1 (370)	0.085±No SE	Heinz et al. (2012)
Elphidium crispum	Yeu Island, France	1 (600)	0.04±No SE	Heinz et al. (2012)
Baculogypsina sphaerulata	Xiao Liu Chiu Island, Taiwan	40 (1,200)	1.95±0.04	Doo et al. (2012)
Marginopora vertebralis	One Tree Island, Australia	60 (600)	171.28±1.37	Doo unpublished data
Calcarina gaudichaudii	Penghu Island, Taiwan	72 (1,440)	16.34±0.46	Doo et al. unpublished data
Calcarina gaudichaudii	Okinawa Island, Japan	60 (600)	8.30±0.25	Doo et al. unpublished data
Amphisorus hemprichii	Okinawa Island, Japan	59 (177)	77.55±2.79	Doo et al. unpublished data

5.2.2 *Extraction of Proteins Using Radioimmunoprecipitation Lysis Buffer*

1. At the termination of the experiment, transfer foraminifera into 1.5 mL microcentrifuge tubes, flash freeze specimens in liquid nitrogen, and store microcentrifuge tubes at −80 °C until extraction of protein.
2. Remove samples from the −80 °C freezer or use freshly isolated samples. Homogenize LBFs in RIPA buffer with protease inhibitor (50 mM Tris–HCl pH [7.4], 1 % Nonidet-P40, 0.25 % Na-deoxycholate, 150 mM NaCl, and 1× complete protease inhibitor cocktail, [Roche, Basel, Switzerland]; Appendix 1). Generally, the optimal volume of RIPA buffer will range from 50 to 200 μL, though this will depend on the number and size of the specimens, but should be approximately 2× the total volume after homogenization. It is important to add the extraction buffer quickly to the sample tubes of frozen foraminifera to prevent protein degradation.
3. Mechanically grind the LBFs in RIPA buffer in 1.5 mL microcentrifuge tubes with metal tweezers until all of the tests are completely pulverized. Alternatively, samples can be ground in a mortar and pestle with liquid nitrogen, and proteins extracted from the resultant powder, though the total protein yield may be lower than with the aforementioned method.
4. Leave the resulting LBF-RIPA lysis buffer solution on ice for 20–30 min and agitate 1–2 times by flicking the tube lightly to resuspend the sample.

5. Centrifuge the homogenate at 12,000 × g for 5 min at 4 °C.
6. Transfer the supernatant to a new 1.5 mL microcentrifuge tube and store it at −80 °C until further analysis. Be careful to remove only the liquid portion, and none of the tests, as inadvertent carry-over of the latter can inhibit downstream analyses.

5.2.3 *Total Protein Quantification*

Total protein quantification is necessary to ensure equal protein loading into the 1-D SDS-PAGE gels. We have found that the detergent-compatible Pierce® BCA Protein Assay kit (Thermo-Scientific #23227) works well for quantification of proteins dissolved in RIPA buffer. Since the detection range for this assay is between 0.2 and 2.0 mg/mL, total protein samples will likely need to be diluted with additional RIPA buffer for analysis of total protein. It is ideal to keep samples at a higher concentration of total protein in order to facilitate more efficient analysis of proteins (e.g., less loading volume). If protein concentrations are below the recommended level of 0.5 mg/mL, an additional TCA precipitation step can be performed to increase the concentration (see Sect. 5.6.3).

1. If samples are stored at −80 °C, defrost them on ice.
2. Turn on the spectrophotometer, and allow it to warm up for at least 30 min dependant on the model.
3. While waiting for the samples to defrost, prepare triplicate samples of each of the bovine serum albumin (BSA) protein standards: 0 mg/mL (ddH_2O only), 0.4 mg/mL, 0.8 mg/mL, 1.2 mg/mL, 1.6 mg/mL, 2.0 mg/mL. Use 10 μL total for each of the triplicate BSA standards.
4. After the homogenates in RIPA buffer are fully defrosted, aliquot 10 μL (or 5 μL of the homogenate and 5 μL of either RIPA buffer or PBS for a 1:1 dilution dependant on initial total protein concentration) in each of triplicate wells of the 96-well plate.
5. Add 200 μL of the "working reagent" (a 50:1 ratio of Reagent A:B) into each well and incubate at 37 °C for 30 min before reading at 562 nm.
6. Create a standard curve between known concentration of BSA (x-axis) and absorbance at 562 nm (y-axis), and calculate the protein concentrations of the LBF samples from their absorbance values. Calculate the mean across the triplicate technical replicates for each sample.

5.3 1-D SDS-PAGE

Equipment and reagents
Heating block (capable of 95 °C)
Microcentrifuge (capable of 4 °C)
Microcentrifuge tubes

(continued)

(continued)

Equipment and reagents
Bio-RAD Mini-PROTEAN® glass plates and casting module
Bio-RAD Mini-PROTEAN® Tetra Cell electrophoresis apparatus
GE healthcare Typhoon™ 9,400 scanner
Laemmli sample buffer
Protein ladder (e.g., Fermentas PageRuler™ Prestained Protein Ladder)
SYPRO® Ruby (Invitrogen)
SDS-PAGE running buffer (see Appendix 1)
Resolving and stacking gel components (see Appendix 1)

5.3.1 Protein Loading Quantities

We have found that in analysis of certain proteins (i.e., the carbon fixation enzyme RuBisCO) successful blots can be performed with as little as 0.125 mg/mL or 5 μg with detection using the SuperSignal® West Pico Chemiluminescent Kit, which increases the signal intensity of the bands by using an enhanced chemiluminescent substrate for detection of horseradish peroxidase (HRP) conjugated with the secondary antibody. It is, therefore, important to utilize an HRP conjugated secondary antibody during the blotting procedure if the SuperSignal® West Pico Chemiluminescent Kit will be used for visualization of bands (see Sect. 5.4.2).

5.3.2 Sample Preparation for SDS-PAGE Gel Electrophoresis

1. The Laemmli sample buffer should normally be prepared without 2-mercaptoethanol for long-term storage. If made in a stock solution, keep at −20 °C and use within 3 weeks.
2. Add a predetermined amount of soluble protein (at least 5 μg) into a new 1.5 mL microcentrifuge tube and dissolve in 5× sample buffer (Laemmli 1970) to a final concentration of 1× sample buffer. Please note that this may require addition of ddH_2O. While preparing samples, keep them on ice to prevent degradation of proteins.
3. Boil all samples for 5 min at 95 °C to denature proteins. Do not heat for more than 5 min, as proteins may degrade.
4. Spin samples at 12,000×g for 10 min at 4 °C to pellet insoluble material, and transport supernatent to a new microcentrifuge tube.
5. Samples can be stored at −80 °C for future analysis or used immediately for electrophoresis.

5.3.3 1-D SDS-PAGE Gel Preparation

Pre-made gels can be bought commercially (e.g., Invitrogen NuPage® Bis-Tris Gel) or made as described below. This protocol will make gel slabs compatible with the Bio-RAD Mini-PROTEAN® Tetrad system.

Table 5.2 Recipes for 12 % resolving and 5 % stacking gels. The recipe below is sufficient for one 1.5 mm acrylamide gel

	% in solution	Total added
A 12 % resolving gel		
ddH_2O	33	3.3 mL
30 % acrylamide/Bis	40	4.0 mL
1.5 M Tris-base, pH 8.8	25	2.5 mL
10 % SDS	1	100 μL
10 % APS	1	100 μL
TEMED	0.04	4 μL
Total volume		10 mL
B 5 % stacking gel		
ddH_2O	68	3.4 mL
30 % acrylamide/Bis	16.6	0.83 mL
1 M Tris, pH 6.8	16.6	0.63 mL
10 % SDS	1	50 μL
10 % APS	1	50 μL
TEMED	0.1	5 μL
Total volume		5 mL

1. Thoroughly clean the flat glass plates, glass spacer plate and corresponding combs (according to the gel thickness used) with 75 % EtOH and dry with paper towels.
2. Assemble the gel-casting chamber as recommended by the manufacturer and check for leaks by pouring ddH_2O into the chamber and observing whether moisture accumulates on the foam pads.
3. After removing the ddH_2O and drying the chamber, determine the volumes needed for the resolving and stacking gels (usually ~10 mL per 1.5 mm mini-gel for one separating gel, see Table 5.2A).
4. Mix the reagents for the resolving gel, adding the TEMED last. After addition of TEMED, quickly vortex the solution, and add it to the gel-casting chamber constructed from the glass plates and the plastic locks from the manufacturer. Pour the gel to a height of approximately 55 mm, or ~18 mm from the top. Please note that acrylamide and TEMED are both extremely hazardous, and the proper precautionary protective measures should be used to protect against contact.
5. Add ~200 μL isopropanol to the top of the gel to flatten it. This will also release bubbles that may have inadvertently formed near the top of the gel. Wait for the gel to solidify (~20–30 min).
6. Meanwhile, prepare the stacking gel without TEMED (Table 5.2B). Please note, the addition of TEMED will cause the gel to solidify, so the TEMED should be added just before pipetting the stacking gel into the gel-casting chamber).
7. After the resolving gel has solidified, discard the isopropanol and wash twice with ddH_2O, discarding the waste each time. Gently dry the gel and glass gel-casting chamber with filter paper.

8. Add the TEMED to the stacking gel solution (see Step 6), and quickly transfer the solution to the gel-casting chamber until almost full.
9. Quickly align and insert the comb into the proper position. Be careful not to trap any air bubbles under the grooves of the comb, as proteins cannot migrate through the resulting gaps.
10. Wait an additional 30–60 min for the stacking gel to dry and then place in 1× running buffer to prevent over-drying (See Appendix 1).
11. Gels can be left in 1× running buffer for 2–3 days at 4 °C.

5.3.4 1-D SDS-PAGE

1. Load a predetermined amount of the protein sample in sample buffer into the gel wells (either handmade in Sect. 5.3.3, or commercially bought, pre-cast gels). The maximum loadable volume will depend on the size of the well, though they typically hold between 15–50 μL. Take care that the sample does not spill over into other lanes.
2. In addition to the samples, each gel should also contain a protein ladder (e.g., Fermentas PageRuler™prestained protein ladder: Cat. No. SM0671), and a positive control protein sample created either in house, or obtained commercially that has been previously shown to work well with the antibody of interest. This will be used to normalize samples between gels.
3. Start the electrophoresis on ice at 70 V for 30 min followed by 120 V for 1 h though the 5 % stacking and 12 % resolving gels, respectively. The exact running times and voltages should be modified to place the target protein at approximately the middle of the gel.

5.3.5 Protein Quality Assessment

Protein samples should be assessed for quality. Degraded protein samples will often demonstrate smearing patterns on the gel and will be of predominantly low molecular weight, while intact proteins will generally be visualized as clear, sharp bands (e.g., Fig. 5.1).

1. Remove the acrylamide gel from the cassette after electrophoresis and discard the stacking gel.
2. Fix the gel by immersing it twice in the fixation buffer (Appendix 1) for 30 min each immersion.
3. Stain the gel with sufficient SYPRO® Ruby (Invitrogen) to cover the gel for 8–12 h at 4 °C in a dark container to prevent light from degrading the reagent.

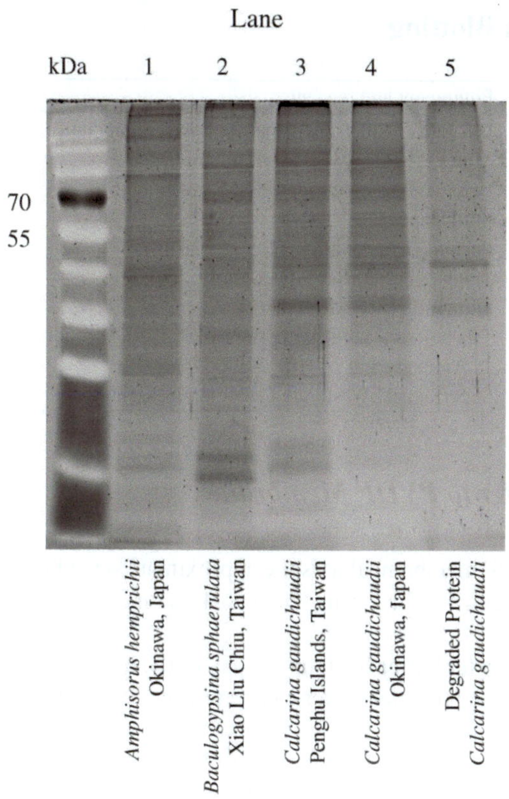

Fig. 5.1 A 1-D SDS PAGE acrylamide gel showing different species of foraminifera that exhibit varied banding patterns. Bands of *Amphisorus hemprichii* (dinoflagellate-bearing species), *Baculogypsina sphaerulata* (diatom-bearing species) and *Calcarina gaudichaudii* (diatom-bearing species) from two locations are shown. Degraded proteins will appear to have a smeared pattern, with the majority of proteins in the lower molecular weight region (Lane 5). A molecular weight marker is shown in the left-most column

Aluminum foil wrapped around a small container works well. Alternatively, a rapid protocol for staining can be used (See Appendix 1).

4. Immerse the gel in wash buffer (Appendix 1) for 30 min.
5. Decant the wash buffer, overlay the gel with ddH_2O and keep in a dark place until visualization.
6. Stained gels can be visualized on a Typhoon Trio™ Variable Mode Imager (Amersham Biosciences, Little Chalfont, United Kingdom) under the SYPRO Ruby (532 nm) setting.

5.4 Western Blotting

Equipment and reagents
Bio-RAD Mini Trans-Blot® Cell, ice cooling unit
Mini gel holder cassette with foam pads
Polyvinylidene fluoride (PVDF) membrane
Filter paper
Fusion FX7 chemiluminescent visualizer
Transfer buffer (See Appendix 1)
Skim milk (See Appendix 1)
Tris-buffered saline with Tween-20® (See Appendix 1)
1° Antibody and 2°-HRP Antibody
SuperSignal® West Pico Chemiluminescent Kit

5.4.1 Transfer to PVDF Membrane

1. Cut out a PVDF membrane that is the approximate size of the mini-gel, and soak it in 100 % methanol for 1 min. Then soak the membrane in 1× transfer buffer for 1 min.
2. Prepare the transfer buffer fresh on the day of the transfer (See Appendix 1), and store at 4 °C. Used transfer buffer can be re-used for the assembly of the blotting cassettes.
3. In a large flat glass or plastic tray, assemble the blotting cassettes with all components (e.g., gel and filter paper) submerged in transfer buffer, making sure that there are no air bubbles between the gel and the PVDF membrane (Fig. 5.2).
4. Electrophorese the gel at 100 V for 90 min at 4 °C, with the ice cooling unit in transfer buffer (25 mM Tris–HCl, pH 6.8, 192 mM glycine, and 20 % methanol).

5.4.2 Blocking and Blotting

1. Carefully remove the PVDF membrane from the transfer cassette and place it into a container with 5 % skim milk (w/v) in Tris-buffered saline with Tween-20 (TBST, 100 mM Tris–HCl, 150 mM NaCl, 0.05 % Tween-20) for 1 h at room temperature (RT).
2. Decant the blocking buffer and replace with 10 mL (or until the PVDF membrane is fully covered) of a 1:500–1:5,000 dilution of the primary antibody (1° antibody) of interest in 5 % skim milk (w/v) in TBST (Table 5.3).
3. Incubate the PVDF membrane in the 1° antibody for 2 h with gentle agitation at RT. Alternatively, samples can be left overnight at 4 °C and gently agitated at RT the following day before continuing the procedure.

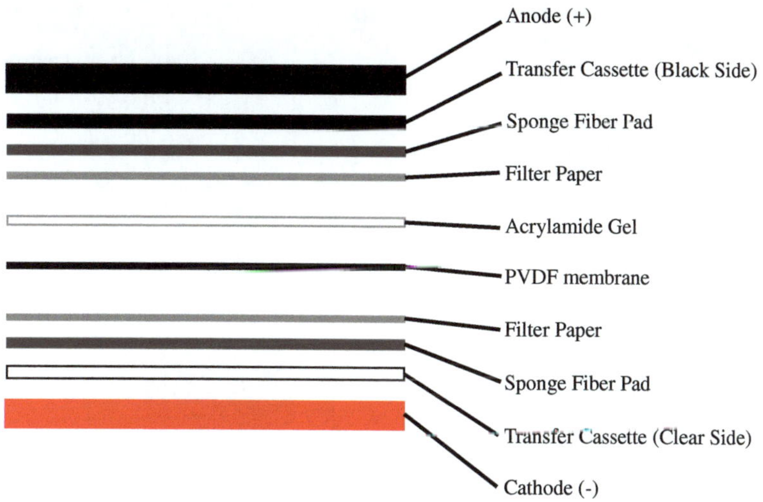

Fig. 5.2 Diagram of electro-blotting cassette in the submerged transfer apparatus. See Sect. 5.4 for further details regarding each component

Table 5.3 Dilutions of antibodies

Dilution	1° antibody
1/500	20 μL in 10 mL TBST buffer w/ 5 % skim milk
1/1,000	10 μL in 10 mL TBST buffer w/ 5 % skim milk
1/5,000	2 μL in 10 mL TBST buffer w/ 5 % skim milk
1/10,000	1 μL in 10 mL TBST buffer w/ 5 % skim milk
	2° antibody
1/5,000	2 μL in 10 mL TBST buffer
1/10,000	1 μL in 10 mL TBST buffer

4. Wash samples three times (5 min each) with TBST, decanting the previously used TBST after each wash.
5. After removing the TBST from the last wash step, incubate the membrane with a 1:5,000 dilution of the compatible secondary antibody (2° antibody) for 1 h with gentle agitation at RT. The secondary antibody used should be HRP conjugated for compatibility with the SuperSignal® West Pico Chemiluminescent Kit.
6. After the 1 h incubation, decant the 2° antibody and wash 3–4 times (5 min each) with TBST.

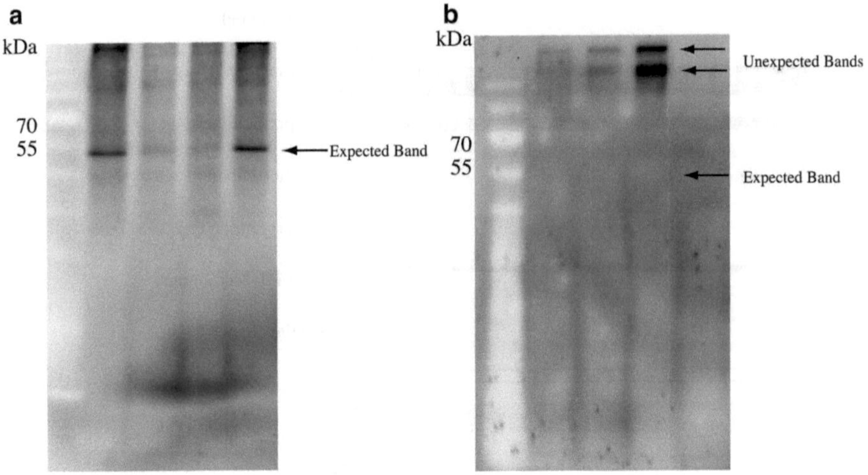

Fig. 5.3 Western blotting of (**a**) RuBisCO protein in the symbiont of *C. gaudichaudii* with clear bands shown at the expected size (~54k Da). (**b**) An unexpected banding pattern resultant from blotting for RuBisCO in *A. hemprichii*

5.4.3 Visualization

We used the chemiluminesent reagent SuperSignal® West Pico Chemiluminescent Substrate to visualize immuno-reactive proteins (outlined below).

1. Mix the two reagents of the SuperSignal® West Pico Chemiluminescent Substrate Kit (Pierce, 34082 Amersham Biosciences) in a 1:1 ratio in an opaque microcentrifuge tube to generate the "ECL" solution.
2. Stain the PVDF membrane with the ECL solution (~400–500 μL), while ensuring that the PVDF membrane is moist at all times. If longer exposure times are needed (greater than 1–2 min), overlay a sheet of transparent plastic over the ECL-coated PVDF membrane to ensure that the chemiluminescent reagent is covering the PVDF membrane at all times.
3. Visualize the blots on a Fusion FX7 (Vilber Lourmat, Marne-la-Vallée, France) imaging doc. Re-expose membranes as needed so that the bands are clearly seen, though not to the extent that the background is over-exposed (Fig. 5.3a). Alternative methods exist using film development in place of a gel doc and can be found in protocols listed by the manufacturer (http://www.piercenet.com/instructions/2160636.pdf).

5.5 Image Analysis

Use Image J (free software from the United States National Institutes of Health) to conduct semi-quantitative densitometry measurements. Normalize each sample of each gel to the intensity of the positive control band of that same gel to compare expression across blots.

5.6 Troubleshooting

5.6.1 Dot Blotting

When testing the reactivity of new antibodies, an initial dot blot test can be performed to provide preliminary evidence of successful reactivity. Because antibody binding is species specific, some antibodies may not bind efficiently to proteins from an organism different from that in which the antibodies were derived. Soak the PVDF membrane in 100 %methanol for >1 min, place a drop of the protein homogenate in RIPA buffer directly onto the PVDF membrane, and allow the protein-coated membrane to dry. Then, follow the standard western blotting steps from the blocking step onwards starting with an initial 1 min wash in transfer buffer. A list of 1° antibodies with known immuno-reactivity to foraminiferal species is listed in Table 5.4.

5.6.2 Unexpected Banding Patterns

Occasionally, unexpected banding patterns may occur; this could result from a variety of factors. For instance, the initial blocking step may have been too short, or the concentration of the 1 or 2° antibodies were too high. In addition, polyclonal antibodies may result in a higher possibility of non-specific binding. It is important to ascertain the cause of this, and apply troubleshooting methodologies to resolve these issues. One common mistake is the incomplete denaturing of the protein in the initial heating step after the addition of the Lamelli buffer (see Sect. 5.3.2). If incomplete denaturing of the secondary protein structures is occurring, smeared bands could be evident. Alternatively, if the protein of interest forms oligomer complexes, this may result in bands of interest at a greater molecular weight than expected. One case of this is in the symbiont (*Symbiodinium* sp.) protein RuBisCO in the foraminifer holobiont, *Amphisorus hemprichii*. The initial heating step performed in this protocol results in multiple non-specific bands, while the band of interest (~54 kDa) is not present (Fig. 5.3b). A longer (~10 min)

Table 5.4 Antibodies with known reactivity to LBFs

Antibody	Compartment	Reference	Manufacturer
RuBisCO Large subunit forms I and II	Symbiont	Doo et al. (2012)	Agrisera, Vännas, Sweden
Hsp70 mouse anti-human MA3-006	Host	Heinz et al. (2012)	Dianova, Hamburg Germany
Hsp70/Hsc70 mouse anti-chicken SPA-822	Host	Heinz et al. (2012)	Dianova, Hamburg Germany

and lower heating temperature (~75 °C) can be used to fully denature more complex protein structures, but troubleshooting of this nature may be necessary on a species-specific level.

5.6.3 *Protein Precipitation Techniques*

When protein concentrations are too low for SDS-PAGE and subsequent western blotting, precipitation can be used.

Equipment and Reagents
1.5 mL microcentrifuge tubes
Low temperature microcentrifuge
100 % (w/v) trichloroacetic acid (TCA)
Acetone

1. Add 1 volume of TCA for every 4 volumes of sample.
2. Incubate for 10 min on ice, occasionally flicking the tube by hand.
3. Spin down the precipitate at 12,000 × g for 10 min at 4 °C.
4. Carefully remove the supernatant without disturbing the white pellet.
5. Wash the pellet with ~200 μL cold acetone and spin at 12,000 × g for 10 min at 4 °C.
6. Carefully decant the acetone and repeat the wash step.
7. Carefully decant the second acetone wash, and leave the sample to dry inverted in a clean hood.
8. When the pellet is dry, add the desired amount of RIPA buffer supplemented with protease inhibitor in order to resuspend the pellet.
9. Store at −80 °C or use immediately in SDS-PAGE after quantification.

5.6.4 *Potential Applications of LBF Protein Analyses*

The ability to probe for proteins of interest in both host and symbiont cells is a major advance for generation of insights into foraminiferal biology and, thus, ecology. Data produced from these methods will contribute to the determination of mechanistic contributions of the host and symbiont to the foraminifer holobiont. By partitioning these two components of the holobiont by targeting host and symbiont specific proteins, details of the overall physiological health can be determined. These analyses performed in parallel with other physiological parameters (i.e., PAM, respiration, *chl-a* concentration, etc.), have the ability to provide novel insights as to why certain species may be more resilient to stress than others, especially in the context of emerging studies on the impacts of climate change on this important taxon.

Appendix 1

	Chemical formula	Final concentration	Amount	Manufacturer
RIPA buffer[a]				
1 M Tris–HCl (pH 6.8)	$NH_2C(CH_2OH)_3$	50 mM	2.5 mL	JT Baker 4099-02
NP-40/Triton X 1 %	$C_{14}H_{22}O(C_2H_4O)_n(n=9\text{-}10)$	1 %	500 μL	Sigma-Aldrich X-100
Sodium-deoxycholate	$NaC_{24}H_{39}O_4$	0.25 %	0.25 g	Thermo Scientific 89904
5M Sodium chloride	NaCl	150 mM	1.5 mL	Sigma-Aldrich 31434
Double-distilled water (ddH_2O)	H_2O		To 50 mL	
Protease inhibitor	Inhibitor Cocktail	per manufacturer's instruction		Roche Complete Protease Inhibitor Cocktail Tablets 04693116001
		Total	50 mL	
2× SDS-PAGE Laemmli sample buffer[b]				
1M Tris–HCl (pH 6.8)	$NH_2C(CH_2OH)_3$	125 mM	1.25 mL	JT Baker 4099-02
10 % SDS	$NaCH_3(CH_2)_{11}OSO_3$	4 % (w/v)	4 mL	JT Baker 4095-02
80 % Glycerol	$HOCH_2CH(OH)CH_2OH$	20 % (w/v)	2.5 mL	Sigma-Aldrich 15523
2 mg/mL Bromophenol blue	$C_{19}H_9Br_4NaO_5S$	0.01 % (w/v)	100 μL	Sigma-Aldrich RD32768
2-Mercaptoethanol	$HSCH_2CH_2OH$	5 %	500 μL	Sigma-Aldrich M7154
ddH_2O	H_2O	–	To 10 mL	–
		Total	10 mL	
10 % SDS[c]				
10 % Sodium dodecyl sulfate (SDS)	$NaCH_3(CH_2)_{11}OSO_3$	10 % (w/v)	25 g	JT Baker 4095-02
ddH_2O	H_2O		To 250 mL	
		Total	250 mL	

(continued)

Appendix 1 (continued)

	Chemical formula	Final concentration	Amount	Manufacturer
10× SDS-PAGE running buffer[d]				
Tris-base	$NH_2C(CH_2OH)_3$	0.25 M	30.3 g	JT Baker 4099-02
Glycine	NH_2CH_2COOH	1.92 M	144.0 g	Sigma-Aldrich G8898
SDS	$NaCH_3(CH_2)_{11}OSO_3$	33 mM	10.0 g	JT Baker 4095-02
ddH_2O	H_2O	–	To 1,000 mL	
		Total	1,000 mL	
1 M Tris–HCl (pH 6.8)[e]				
Tris-base	$NH_2C(CH_2OH)_3$	1 M	60.57 g	JT Baker 4099-02
12 N HCl	HCl	–	varies	
ddH_2O	H_2O	–	To 500 mL	
		Total	500 mL	
1.5 M Tris–HCl (pH 8.8)[f]				
Tris-base	$NH_2C(CH_2OH)_3$	1.5 M	90.86 g	JT Baker 4099-02
12N NaOH	NaOH	–	varies	
ddH_2O	H_2O	–	To 500 mL	
		Total	500 mL	
10 % APS[g]				
Ammonium persulfate (APS)	$(NH_4)_2S_2O_8$	10 % (w/v)	0.5 g	JT Baker 0762-01
ddH_2O	H_2O	–	To 5 mL	
		Total	5 mL	
Bromophenol blue[h]				
Bromophenol blue	$C_{19}H_9Br_4NaO_5S$	2 mg/mL	10 mg	Sigma-Aldrich RD32768
ddH_2O	H_2O	–	To 5 mL	
		Total	5 mL	

Gel Fixation Buffer prior to SYPRO® Ruby Stain[i]				
Methanol	CH_3OH	50 % (v/v)	500 mL	Sigma-Aldrich 34966
Acetic acid	CH_3COOH	7 % (v/v)	70 mL	Sigma-Aldrich A6283
ddH_2O	H_2O	–	To 1,000 mL	
		Total	1,000 mL	
Destaining buffer for SYPRO® Ruby Stain				
Methanol	CH_3OH	10 % (v/v)	100 mL	Sigma-Aldrich 34966
Acetic acid	CH_3COOH	7 % (v/v)	70 mL	Sigma-Aldrich A6283
ddH_2O	H_2O	–	To 1,000 mL	
		Total	1,000 mL	
5× SDS-PAGE transfer buffer[j]				
Tris-base	$NH_2C(CH_2OH)_3$	0.123 M	15.0 g	JT Baker 4099-02
Glycine	NH_2CH_2COOH	0.959 M	72.0 g	Sigma-Aldrich G8898
ddH_2O	H_2O	–	Fill to 1,000 mL	
		Total	1,000 mL	
1× SDS-PAGE transfer buffer[k]				
Methanol	CH_3OH	20 % (v/v)	200 mL	Sigma-Aldrich 34966
5× transfer buffer	–	20 % (v/v)	200 mL	–
ddH_2O	H_2O	–	To 1,000 mL	
		Total	1,000 mL	
10× TBS buffer[l]				
Tris-base	$NH_2C(CH_2OH)_3$	0.1M	12.11 g	JT Baker 4099-02
Sodium chloride	NaCl	1.5M	87.66 g	Sigma-Aldrich 31434
12N HCl	–	–	varies	
ddH_2O	H_2O	–	To 1,000 mL	
		Total	1,000 mL	

(continued)

Appendix 1 (continued)

	Chemical formula	Final concentration	Amount	Manufacturer
1× TBST Buffer[m]				
10× TBS	–	10 % (v/v)	100 mL	–
Tween-20	$C_{58}H_{114}O_{26}$	0.5 % (v/v)	500 μL	Sigma-Aldrich P5927
ddH_2O	H_2O	–	To 1,000 mL	–
		Total	1,000 mL	
5 % blocking buffer				
1× TBST buffer	–	–	10 mL	
Skim milk	–	5 % (w/v)	0.5 g	Sigma-Aldrich Fluka 70166
		Total	10 mL	

[a]Instructions: Add the protease inhibitor right before use, as protease inhibitors will degrade upon continual storage in liquid form. Store RIPA lysis buffer at 4 °C
[b]Instructions: Add all components and keep at −20 °C for use within 3 weeks. Add 2-Mercaptoethanol in a fumehood, and avoid repeated freeze-thaws
[c]Instructions: Dissolve 25 g SDS in 200 mL ddH2O, then fill to 250 mL
[d]Instructions: Dissolve Tris-base, glycine and SDS in 900 mL ddH_2O, then fill to 1,000 mL. Dilute to 1× with ddH_2O before use in gel electrophoresis
[e]Instructions: Dissolve Tris-base in 400 mL ddH_2O, adjust pH to 6.8 with 12 N HCl and fill to 500 mL with ddH_2O
[f]Instructions: Dissolve Tris-base in 400 mL ddH_2O, adjust pH to 8.8 with 12N NaOH, and fill to 500 mL with ddH_2O
[g]Instructions: Dissolve APS powder in 5 mL ddH_2O. Aliquot into 1.5 mL eppendorff tubes, and store at −20 °C for a maximum of 4 weeks. Avoid repeated freeze-thaws
[h]Instructions: Dissolve bromophenol blue in 5 mL ddH_2O. Aliquot into 1.5 mL eppendorff tubes, and store at −20 °C
[i]Instructions: Add and mix chemicals in the fume hood. Rapid Protocol: (1) Microwave gel for 30 s, agitate (70 rpm) for 30 s. (2) Microwave gel for 30 s, agitate (70 rpm) for 5 min. (3) Microwave gel for 30 s, agitate (70 rpm) for 25 min. (4) Continue with Step 4 in Sect. 5.3.5
[j]Instructions: Dissolve Tris-base and glycine in 900 mL ddH_2O, then fill to 1,000 mL
[k]Instructions: Make up beforehand, and store at 4 °C
[l]Instructions: Dissolve tris-base and sodium chloride in 900 mL ddH_2O, adjust pH with 12 N HCl to pH 7.6, and fill to 1,000 mL
[m]Instructions: Mix TBS, ddH_2O and Tween-20, and mix well

References

Doo SS, Mayfield AB, Byrne M, Chen H-K, Nguyen HD, Fan T-Y (2012) Reduced expression of the rate-limiting carbon fixation enzyme RuBisCO in the benthic foraminifer *Baculogypsina sphaerulata* holobiont in response to heat shock. J Exp Mar Biol Ecol 430–431:63–67

Fujita K, Fujimura H (2008) Organic and inorganic carbon production by algal symbiont-bearing foraminifera on northwest Pacific coral-reef flats. J Foramin Res 38:117–126

Glas MS, Fabricius KE, De Beer D, Uthicke S (2012) The O_2, pH and Ca^{2+} microenvironment of benthic foraminifera in a high CO_2 World. PLoS ONE 7:e50010

Heinz P, Marten RA, Linshy VN, Haap T, Geslin E, Köhler H (2012) 70 kD stress protein (Hsp70) analysis in living shallow-water benthic foraminifera. Mar Biol Res 8:37–41

Hikami M, Ushie H, Irie T, Fujita K, Kuroyanagi A, Sakai K, Nojiri Y, Suzuki A, Kawahata H (2011) Contrasting calcification responses to ocean acidification between two reef foraminifers harboring different algal symbionts. Geophys Res Lett 38, L19601

Laemmli UK (1970) Cleavage of structural proteins during the assembly of the head of bacteriophage T4. Nature 227:680–685

Langer MR (2008) Assessing the contribution of foraminiferan protists to global ocean carbonate production. J Eukaryot Microbiol 55:163–169

Lee JJ (2006) Review article algal symbiosis in larger foraminifera. Symbiosis 42:63–75

Rink S, Kühl M, Bijma J, Spero HJ (1998) Microsensor studies of photosynthesis and respiration in the symbiotic foraminifer *Orbulina universa*. Mar Biol 131:583–595

Schmidt C, Heinz P, Kucera M, Uthicke S (2011) Temperature-induced stress leads to bleaching in larger benthic foraminifera hosting endosymbiotic diatoms. Limnol Oceanogr 56:1587–1602

Sinutok S, Hill R, Doblin MA, Wuhrer R, Ralph PJ (2011) Warmer more acidic conditions cause decreased productivity and calcification in subtropical coral reef sediment-dwelling calcifiers. Limnol Oceanogr 56:1200–1212

Yamamoto S, Kayanne H, Terai M, Watanabe A, Kato K, Negishi A, Nozaki K (2012) Threshold of carbonate saturation state determined by CO_2 control experiment. Biogeosciences 9:1441–1450

Chapter 6
Molecular Assessment of Benthic Foraminiferal Diversity

Béatrice Lecroq

Abstract Recent advances in molecular biology have extended our possibilities to investigate biodiversity and evolution of small organisms. This is especially true for the hyper-diverse benthic foraminifera, which can now be identified by their DNA sequences rather than only by their morphology trough time-consuming observations under the microscope. In this chapter, we will introduce two general methods for the molecular assessment of benthic foraminiferal diversity. The first one is "the single-cell approach" and consists in isolating a single living specimen before extracting, amplifying, cloning and sequencing its DNA. The second method is "the environmental DNA approach", which aims to extract the DNA from the overall fauna present in the sediment before specifically amplifying foraminiferal fragments and massively sequence them. For both approaches, we will present some core experimental protocols with specificities and possible optimizations.

Keywords Polymerase chain reaction (PCR) • Cloning • Environmental DNA • High throughput sequencing (HTS)

6.1 Introduction

Since the end of the 1990s, molecular tools have opened new horizons in foraminiferal research by broadening investigations of their biology and environmental diversity through the analysis of DNA sequences (Langer et al. 1993; Pawlowski et al. 1995). Diverse sequence fragments of the foraminiferal ribosomal RNA gene (rDNA) can now be used to confirm the identity of a specimen (Pawlowski 2000) or distinguish

B. Lecroq (✉)
Okinawa Institute of Science and Technology, 1919-1 Tancha,
Onna-son, Kunigami-gun, Okinawa 904-0495, Japan
e-mail: beatrice.lecroq@oist.jp

H. Kitazato and J.M. Bernhard (eds.), *Approaches to Study Living Foraminifera: Collection, Maintenance and Experimentation*, Environmental Science and Engineering,
DOI 10.1007/978-4-431-54388-6_6, © Springer Japan 2014

between cryptic species (Morard et al. 2009). Phylogenetic inferences are particularly conclusive owing to the large amount of evolutionary information accumulated in the DNA sequences of foraminifera. Within the powerful framework offered by molecular phylogenetics, the classification of the group within higher taxa has been resolved (Pawlowski et al. 1994; Pawlowski and Burki 2009), and both the dispersion of species and the resulting gene flow (Lecroq et al. 2009) as well as their occurrence in unexpected habitats such as rivers and soils (Holzmann et al. 2003; Lejzerowicz et al. 2010) have been evidenced. Moreover, molecular methods offer the opportunity to highlight and study large benthic foraminiferal squatters and endosymbionts that often escape traditional microscopic observations (e.g., Pawlowski et al. 2001; Holzmann et al. 2006). Relying on powerful identification techniques is a crucial issue to understand their metabolism and to prevent misinterpretations of "living assemblages" obtained from Rose Bengal-stained sediment samples.

High throughput sequencing (HTS) technologies have revolutionized our view of foraminiferal species richness and diversity by enhancing the screening of environmental samples. Numerous novel lineages that were yet escaping from conventional micropaleontological observations have been revealed. The massive sequencing of environmental DNA from deep-sea sediment has shown high proportions of unknown monothalamous (single-chambered) taxa; challenging our traditional perception of foraminiferal diversity dominated by multi-chambered species (Lecroq et al. 2011). HTS applied to environmental DNA is a particularly relevant approach to tackle inconspicuous groups of benthic foraminifera that could be small, that could lack distinctive features or that could be difficult to culture and clone. From a broader point of view, this approach has simultaneously reduced the cost and time required to assess the taxonomic diversity and opened an analytical way free of culturing and cloning biases.

The estimation of benthic foraminiferal diversity using DNA sequencing methods mainly relies on the reliability of the reference sequence database. The two main criteria used to define a good database are its size and quality, which in turn depend both on our general knowledge of foraminiferal richness and on our ability to faithfully describe or annotate specimens isolated from by microscopic observations and culturing experiments. Single-cell approaches conducted under careful laboratory settings are mandatory to link a marker sequence to a taxon. Indeed, only sequences originating from unique specimens unambiguously associated with a species, or at least a genus name must be allowed to constitute a new entry in the reference database. The alternative of sequence database cleansing is a tedious task, particularly when environmental sequences representing non-described species or artifacts are considered (Berney et al. 2004). So far, most of the molecular studies on foraminifera were based on the small-subunit of the ribosomal RNA gene (SSU rDNA). However, the choice of the molecular markers to be employed is important since the existence of a universal marker performing equally across all branches of such a diverse tree is very unlikely. Contrasting taxonomic resolutions have been reported for the foraminiferal SSU rDNA, with various regions being optimal to sub-optimal at delineating species depending on the analyzed taxa (Pawlowski and Lecroq 2010). Therefore, different markers of satisfying resolution might have to be considered to fully access the complexity of foraminiferal assemblages occurring in environmental samples. Finally, molecular-marker characteristics (especially their size) should also be chosen according to the requirements of the analysis method.

Some of the likely future applications of foraminiferal diversity research, such as baseline and impact environmental studies for deep-sea mining, will require rapid and over time monitoring of the foraminiferal composition in the sediment. Recent molecular techniques have more than ever the potential to answer those needs. However, we are still far from handling and fully exploiting this powerful tool. More studies including replicates, standards and blank references are needed for each step of environmental DNA-approach methods to apprehend their pitfalls and enable accurate and meaningful interpretations of results. This chapter does not aim to display most of the molecular techniques and protocols available today but, rather, to provide a general basic path from where to start to build up any molecular protocol to assess benthic foraminiferal diversity.

6.2 Single-Cell Approach

6.2.1 Preparation

The main objective of sampling in situ foraminifera in the single-cell approach is to isolate and preserve the cytoplasm of a single specimen. To avoid rapid degradation of DNA, all the preliminary manipulations of sampling material (sieving sediment, brushing specimens from rocks or rubble, observation under microscope and sorting) should be performed as rapidly as possible after recovering in controlled temperature and light conditions. In case of sieving and during sorting the filtered water from the collection site should be used. Sediment can be stored at 4 °C for a couple of days (better to separate fractions before) or conserved raw and frozen at −80 °C or −20 °C for long conservation.

- *Isolation:*
  The key point of this sampling step is to distinguish live specimens (containing nuclei and DNA) from dead ones. Specimen with visible and colorful cytoplasm (greenish, brownish, whitish) will be preferentially selected. For species with non-transparent test it is sometime necessary to carefully break the test to check for the presence of cytoplasm. In that case, cytoplasmic material should be sucked up with a pipette before putting it immediately into extraction buffer. Alternatively, it is possible to select living specimen by observing their cytoplasmic activity outside the test. The stress of the sampling and sorting manipulations often causes the reticulopodia to retract. However, a living specimen will probably extend its pseudopodia again if kept without perturbation in the refrigerator for some hours. Isolated specimens (including deep-sea specimens) can be maintained alive in Petri dishes in the refrigerator for a couple of days without any manipulation. Staining solutions should be avoided since they usually prevent PCR amplification. Before putting the specimen or its cytoplasm in DNA extraction buffer, sample must be cleaned of any DNA from other organisms (especially from other foraminiferal specimen damaged during sieving for instance).

- *Storage:*
  Isolated specimen can be conserved frozen (better using liquid nitrogen before storing) at −80 °C for long conservation or at −20 °C for some years. For calcareous and agglutinated hard-shelled species, air-dried samples can be stored at room temperature for a couple of years. Foraminiferal specimens conserved in ethanol or formaldehyde are generally not suitable for molecular analysis since those chemical fixatives and the sub-products of their degradation may inhibit DNA amplification by PCR. Specimens can also be conserved in extraction buffer such as guanidine (for DNA only) or RNAlater™ (for DNA and RNA) at room temperature up to a week, at 4 °C for some days or at −30 °C to −20 °C for some years.

6.2.2 DNA Extraction

DNA extraction from single cell can be performed with commercially-available kits but we rather recommend the following protocol using guanidinium isothiocyanate "guanidine" buffer and modified from Clark (1992).

- *Guanidine preparation protocol:*

  1. To 100 g of guanidinium isothiocyanate, add deionized 100 mL of H_2O, 10.6 mL of 1M TRIS/HCl pH 7.6, 10.6 mL of 0.2M Ethylenediaminetetraacetic acid "EDTA".
  2. Stir overnight at room temperature (heat to 60–70 °C for 10 min to assist dissolution).
  3. Add 21.2 mL of 20 % Na lauryl sarkosinate "Sarkosyl™".
  4. Bring the volume to 210 mL with sterile H_2O and filter.
  5. Add 2.1 mL of β-mercaptoethanol and mix.
  6. Store at 4 °C in a brown glass bottle.

- *DNA extraction protocol:*

  1. Place the specimen in 60 μL of guanidine buffer and vortex briefly.
  2. Heat 10 min at 60 °C.
  3. Centrifuge at 10,000 rpm for 2 min.
  4. Add 50 μL of the supernatant to 50 μL of isopropanol and vortex briefly.
  5. Precipitate overnight (or at least 2 h at −20 °C).
  6. Centrifuge at 15,000 rpm for 15 min at 5 °C.
  7. Discard the supernatant and carefully wash the pellet with 300 μL of 70 % ethanol.
  8. Centrifuge at 15,000 rpm for 5 min.
  9. Discard the supernatant and allow ethanol to dry from the pellet.
  10. Gently dissolve in 50 μL of sterile H_20 (keep on ice).

Best results are obtained by grinding specimens with a sterile pestle grinder. However, if the foraminiferal test should be conserved for morphological studies, it is possible to use an EDTA-free guanidine buffer.

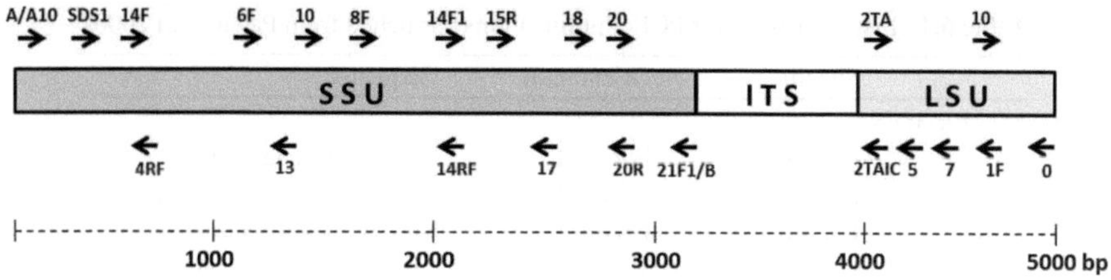

Fig. 6.1 Schematic representation of the small-subunit (SSU), the internal transcripted spacer (ITS) and the large-subunit of the ribosomal DNA (rDNA) sequence in foraminifera (retouched from Pawlowski 2000)

One single specimen of size >100 μm diameter contains generally enough DNA to observe bands on electrophoresis gel after PCR (nested PCR, 30+30 cycles or less). For smaller specimens (30–100 μm diameter), up to ten specimens may be required. It is noteworthy that significant numbers of sediment particles can be loaded into the extraction buffer in case of agglutinated species; particle presence could impact results. Sediment particles often contain inhibitors that could prevent PCR amplification. Thus, in the case of large agglutinated species, it is preferable to break the agglutinated test and add only the cytoplasm to guanidine extraction buffer.

6.2.3 DNA Amplification

Targeted regions of the foraminiferal SSU rDNA marker could be a enriched from diverse single-cell or environmental DNA extracts by standard PCR amplification. Specific foraminiferal primers have been designed to anneal to positions of the SSU rDNA that are specific to foraminifera (Pawlowski et al. 1994). Multiple primers pairs can be tested as well as in combination with "universal", eukaryotic primers, allowing coverage of the full ribosomal gene. The foraminiferal SSU rDNA gene sequence is unusually long. Although long-range PCR are possible, several PCR rounds are often necessary to retrieve the complete gene. Nested PCR is often required in order to obtain enough material from single-specimen extracts and better enrich the targeted foraminiferal DNA while preventing the co-amplification of non-foraminiferal DNA originating either from squatters living inside the shells or symbionts (Fig. 6.1 and Table 6.1).

Strict laboratory procedures are of utmost importance for DNA contamination to be avoided. PCR products are the most likely source of contaminants and, in the case of the SSU rDNA, might promote the formation of chimeric sequences. Indeed, the SSU rDNA gene is interspersed with highly conserved domains where two different templates might hybridize during the PCR and form a chimera "break point". Trace-contaminating remnants from previous DNA amplifications might induce chimera formation, especially as a result of using low annealing temperatures during PCR cycles (53 °C). Therefore, it is recommended to carry out the

Table 6.1 List of primers for PCR amplification (retouched from Pawlowski 2000)

	Specificity	Orientation	Sequence
SSU PRIMER			
SA	Universal	Forward	ggt tga t(ct)c tgc cag a
SA10	Forams	Forward	ctc aaa gat taa gcc atg caa gtg g
SDS1	Forams	Forward	gtt tgg cta ata cgt acg
S4F	Forams	Forward	tct aag gaa cgc agc agg
S4rf	Forams	Reverse	cgc ctg ctg cgt tcc tta g
S6f	Forams	Forward	ccg cgg taa tac cag ctc
S13	Forams	Reverse	gca aca atg att gta tag gc
S10	Forams	Forward	cac tgt gaa caa atc ag
S8f	Forams	Forward	tcg atg ggg ata gtt gg
S14rf	Forams	Reverse	cct tca agt ttc aca ctt gc
S14F1	Forams	Forward	aag ggc acc aca aga acg c
S15r	Forams	Forward	gtg gtg cat ggc cgt
S17	Forams	Reverse	cgg tca cgt tcg ttg c
S18	Universal	Forward	taa cag gtc tgt gat gcc
S20	Universal	Forward	ttg tac aca ccg ccc gtc
S20r	Universal	Reverse	gac ggg cgg tgt gta caa
S21F1	Forams	Reverse	cct tgt tac gac ttc tc
Rib B	Universal	Reverse	tga tcc ttc tgc agg ttc acc tac
LSU PRIMER			
2TA	Forams	Forward	cac atc agc tcg agt gag
2TAIC	Forams	Reverse	ctc act cga gct gat gtg
L5	Universal	Reverse	ttc (ag)ct gcg c(ag)t tac t
L7	Forams	Forward	gat g(at)g tca tta cca cc
L1F	Forams	Forward	act ctc tct ttc act cc
L10	Universal	Forward	ctg acg tgc aaa tcg tt
LO	Universal	Reverse	gct atc ctg ag(ag) gaa act tcg

DNA extraction and PCR mix preparation on different laboratories / rooms. It is also highly recommended to store DNA extracts, PCR reagents and PCR products in different locations. Nevertheless, a thorough verification of the sequenced PCR products is necessary, and potential chimeras should be investigated manually based on permissive chimera detection thresholds.

– *PCR reaction mix:*

MIX	1x (μL)
ΔH_2O	41
Template	2
10x buffer	5
dNTPs	1
Primer forward	1
Primer reverse	1
Enzyme (TAQ)	0.2

– *PCR amplification protocol:*

(1) Preheating	95 °C	5 min	
(2) Denaturation	94 °C	1 min	
(3) Primers annealing	53 °C	1 min	20–40 cycles
(4) Primers extension	72 °C	2 min	
(5) Final extension	72 °C	7 min	

In addition to the choice of the primers, the performance of the PCR enrichment should be optimized. According to our previous experience with foraminiferal DNA, we observed that two parameters are mainly responsible for the success of the amplification of the SSU rDNA fragments: (1) the quantity of template DNA to be added to the reaction volume and (2) the number of PCR cycles and PCR rounds (nested PCR are more susceptible to cross-contamination). If the electrophoresis gel fails to display any visible bands after 40 cycles of PCR, 2-rounds of nested PCR (30 + 25 cycles) should be performed. PCR inhibitors might be co-extracted with the targeted genomic DNA (especially from bulk, organic-rich sediments). The action of inhibitors during PCR can be mitigated by diluting DNA extraction or by appending between 0.4 and 0.8 g L^{-1} of BSA in the PCR mix. Purified PCR products can be conserved frozen at −30 °C for some years. Low-binding polyallomer tubes are preferable for long-term storage of short and/or rare DNA templates.

6.2.4 DNA Cloning and Sequencing

- *Cloning:*
  For the molecular analysis of a single foraminiferal cell, cloning is required for two reasons: (1) to increase the number of DNA sequences and (2) to separate sequences from contaminants or endosymbionts and from the intraindividual polymorphism of rDNA copies (Holzmann et al. 1996). Rotaliids in particular are known to display high variability in the ITS region.
- *Sequencing:*
  Some foraminiferal sequences are known to be especially rich in AT and may contain long series of *poly-A* that hinders sequencing and require a special mix of dNTPs for the sequencing reaction.

6.3 Environmental DNA Approach

The environmental DNA approach or "metagenetics" consists in getting numerous sequences of the global DNA extractions from environmental samples. This strategy allows specialists studying different taxa to share samples since global DNA extractions can be amplified with different specific primers targeting distinct taxonomic groups.

DNA extraction captures environmental metagenomes, including intracellular, coding and non-coding DNA from all species that could be disrupted during lysis. It also captures extracellular DNA (eDNA) material either actively released during cell life (Vlassov et al. 2007) or post-mortem (Haile et al. 2007; Corinaldesi et al. 2007). Identifying the foraminiferal species of an assemblage by assessing its entire environmental DNA content represents an appealing alternative to the time-consuming microscopic observation of each individual morphology. The discovery of DNA sequences belonging to planktonic taxa in deep-sea sediments highlighted the fact that eDNA could be preserved in the sediment long after cell death (Pawlowski et al. 2011; Lejzerowicz et al. 2013). Nevertheless, environmental DNA approaches remain particularly relevant to assess foraminiferal diversity from deep-sea sediments where monothalamous taxa are dominant. For the purpose of identifying these inconspicuous foraminifera, metagenetics outperforms the usually time-consuming steps of isolation and observation under light microscope before morphological identification. Finally, this approach could be applied proficiently for generating large datasets for biomonitoring surveys using foraminifera as ecotoxicological markers and by offering, at a reasonable cost and within a short time, a snapshot of the species composition.

6.3.1 Preparation

The high throughput provided by recent sequencing technologies requires special care to prevent both extraneous and cross-contamination within environmental DNA samples. Moreover, since both the cytoplasmic material and the sedimentary eDNA are indifferently collected during the extraction process and since eDNA has been shown to be particularly abundant in marine sediments (Dell'Anno and Danovaro 2005), the number of template sequences stemming from undesirable contaminating or extracellular DNA might average the number of sequences corresponding to the actual biocoenosis DNA. For the very same reason, it is important to replicate the sampling of a location to increase chances of distinguishing targeted sequences (from living in situ specimens) from others.

Sediments should be immediately frozen after recovery, preferentially by total immersion of a cryo vial filled with the material in liquid nitrogen and then stored at −80 °C for extended periods of time. Deep freezing guarantees the preservation of RNA molecules, which could be used as template for reverse-transcription of both the rRNA and mRNA transcripts. This allows the study of environmental metatranscriptomes as well as it supports unambiguous identification of the species that are indeed living and active at the time of sampling (Lejzerowicz et al. 2013). However, it remains difficult to estimate how long sediment samples can be conserved for further molecular analyses and whether biochemical modifications on the primary structure of DNA molecules operate over time.

6.3.2 Environmental DNA Extraction

Once again, special care is required to avoid contamination during the extraction step. It is recommended to perform the extraction protocol on a sterile, DNA decontaminated bench or under a dedicated hood. The best practice is to perform extractions in a remote laboratory with no history of foraminiferal work, or at least where no single-cell DNA extraction or PCR amplification have been done. The samples should be thawed on ice and put, as soon as possible, into extraction buffer.

Several environmental DNA extraction protocols are available in the literature (Habura et al. 2004). Several DNA extraction kits for soils are also available and offer higher reproducibility of the method. For instance, the *MoBio* kits have been successfully used to extract and massively sequence foraminiferal DNA from the sediment; *MoBio PowerSoil*™ kit for small samples (up to 0.5 g of sediment); or *MoBio PowerMax Soil*™ kit for bigger volumes (up to 5 g).

In case of watery sediment it is recommended to perform a short centrifugation to remove the maximum amount of water before extraction. This manipulation will also contribute to standardize the samples regarding sediment types for further comparison between locations.

It is also possible to extract both DNA and RNA from the same sediment samples (*MoBio RNA PowerSoil*™ *Total RNA Isolation Kit + RNA PowerSoil*™ *DNA Elution Accessory Kit*). In that case, RNA extraction should be treated with DNAase in order to remove the co-extracted DNA molecules before generating the complementary stand (cDNA) by performing the reverse-transcription of the RNA, single-stranded molecules.

DNA extraction products from sediments can be conserved at −30 °C to −20 °C for some years, although we lack experience to predict chemical degradation during long-term conservation for environmental DNA samples.

6.3.3 DNA Amplification, Sample Multiplex and **Illumina, *High-throughput Sequencing***

The general path for the environmental DNA approach is:

1. DNA extractions
2. PCR amplifications (with foraminiferal specific primers)
3. Parallels PCR amplifications with multiplex tags
4. Pooling of tagged PCR products
5. PCR amplifications with sequencing adaptors
6. Massive sequencing
7. High-throughput data analysis

Step 3 appends unique barcoding tag sequences (usually 5–8 nucleotides per tag) to each sample of different origins. The PCR products resulting from step 3 will then be quantified and pooled together in equimolar ratio (or according to the configuration satisfying a requirement for a minimum number of sequences desired per sample). Step 5 attaches to the tagged PCR products some adapters that will connect sample sequences to sequencing support surface (beads in emulsion or flow cell). Steps 3–6 are specific to the chosen sequencing technique and are usually performed according to the commercially available library preparation kits. Therefore, we will not present or recommend any protocol for those steps. Alternatively, the full length adapters could be directly appended to the target fragments by designing long primer constructs also including the tag sequences (Kircher et al. 2011).

Protocols and general recommendations for the first PCR amplifications of extracted DNA are the same as in the "single-cell approach" (Sect. 6.2.3) except that samples contain a mix of many different DNA sequences, that initial foraminiferal DNA concentrations are much lower and inhibitor concentrations may be much higher depending on the sediment type. In order to amplify a maximum of different foraminiferal sequences, the annealing temperature can be slightly lower (50 °C for instance) and the numbers of cycles can be reduced while the number of rounds can be increased (for instance 3–4 rounds of 5–15 cycles). Finally, it is recommended to perform replicates of PCR amplifications from the same DNA extraction that can be pooled later.

6.3.4 Massive Sequencing

We will not give an exhaustive review of all the massive parallel sequencing technologies available, since machine performances are evolving very fast. However, here are the major technologies that could be applied to foraminiferal diversity assessment (number of DNA sequence per run and their maximal size are indicated within brackets): *Roche 454* (100,000 reads of 400 bp), *GS FLX Titanium* (500 reads of 1,000 bp), *Illumina* (5,000,000 reads of 250 bp), *Life Technologies Solid 4* (5,000,000 reads of 75 bp).

Pawlowski and Lecroq (2010) have analyzed six variable regions situated in the most commonly studied fragment of the SSU rDNA looking for a good barcode for foraminifera. They have found that a 36 bp-long fragment located at the foraminiferal specific helix 37f can be used to identify most of the taxa to the genus level. With increasing performance of HTS technologies, the restrictive requirement of such a short barcode has disappeared. However, methods focusing on rather short reads and high number of reads per run such as *Illumina* remain advantageous for foraminiferal diversity assessments.

Finally, additional blanks, standards, DNA extraction from monoclonal cultures and a known foraminiferal assemblage of reference species need to be included in the massive sequencing run.

6.3.5 High-throughput Data Analysis

Analyzing the numerous reads obtained by massive sequencing are probably in the biggest challenge of the "environmental DNA approach". Such analysis consists basically in three main steps: (1) reads filtering, which is based mainly on their quality and occurrence; (2) sequence clustering, which group similar sequences together into organizational taxonomic unit (OTU); and (3) taxonomic identification. So far, this process is performed by customized pipeline programming. The clustering, the taxonomic identification and the program itself that will perform clustering and identification of the reads needs to be customized according to the read size and quality, the blanks and references results, the molecular marker, the taxonomic resolution targeted and the data base of reference.

References

Berney C, Fahrni J, Pawlowski J (2004) How many novel eukaryotic "kingdoms"? Pitfalls and limitations of environmental DNA surveys. BMC Biol 2:13

Clark CG (1992) DNA purification from polysaccharide-rich cells. In: Protocols in protozoology, vol 1. Allen, Lawrence, p D-3.1

Corinaldesi C, Dell'Anno A, Danovaro R (2007) Early diagenesis and trophic role of extracellular DNA in different benthic ecosystems. Limnol Oceanogr 52:1710–1717

Dell'Anno A, Danovaro R (2005) Extracellular DNA plays a key role in deep-sea ecosystem functioning. Science 309:2179

Habura A, Pawlowski J, Hanes SD, Bowser SS (2004) Unexpected foraminiferal diversity revealed by small-subunit rDNA analysis of Antarctic sediment. J Eukaryot Microbiol 51(2):173–179

Haile J, Holdaway R, Oliver K, Bunce M, Gilbert TP, Nielsen R, Munch K, Ho SYW, Shapiro B, Willerslev E (2007) Ancient DNA chronology within sediment deposits: are paleobiological reconstructions possible and is DNA leaching a factor? Mol Biol Evol 24:982–989

Holzmann M, Piller W, Pawlowski J (1996) Sequence variations in large-subunit ribosomal RNA gene of Ammonia (Foraminifera, Protozoa) and their evolutionary implications. J Mol Evol 43:145–151

Holzmann M, Habura A, Giles H, Bowser SS, Pawlowski J (2003) Freshwater foraminiferans revealed by analysis of environmental DNA samples. J Eukaryot Microbiol 50(2):135–139

Holzmann M, Berney C, Hohenegger J (2006) Molecular identification of diatom endosymbionts in nummulitid Foraminifera. Symbiosis 42:93–101

Kircher M, Sawyer S, Meyer M (2011) Double indexing overcomes inaccuracies in multiplex sequencing on the Illumina platform. Nucleic Acids Res 40:1–8

Langer MR, Lipps JH, Piller WE (1993) Molecular paleobiology of protists: amplification and direct sequencing of foraminiferal DNA. Micropaleontology 391:63–68

Lecroq B, Gooday AJ, Pawlowski J (2009) Global genetic homogeneity in deep-sea foraminiferan Epistominella exigua (Rotaliida: Pseudoparrellidae). Zootaxa 2096:23–32

Lecroq B, Lejzerowicz F, Bachar D, Christen R, Esling P, Baerlocher L, Østerås M, Farinelli L, Pawlowski J (2011) Ultra-deep sequencing of foraminiferal microbarcodes unveils hidden richness of early monothalamous lineages in deep-sea sediments. Proc Natl Acad Sci USA 108(32):13177–13182

Lejzerowicz F, Pawlowski J, Fraissinet-Tachet L, Marmeisse R (2010) Molecular evidence for widespread occurrence of Foraminifera in soils. Environ Microbiol 12:2518–2526

Lejzerowicz F, Esling P, Majewski W, Szczucinski W, Decelle J, Obadia C, Arbizu PM, Pawlowski J (2013) Ancient DNA complements microfossil record in deep-sea subsurface sediments. Biol Lett 9(4):20130283

Morard R, Quillévéré F, Escarguel G, Ujiie Y, de Garidel-Thoron T, de Vargas C (2009) Morphological recognition of cryptic species in the planktonic foraminifer Orbulina universa. Micropaleontology 71:148–165

Pawlowski J (2000) Introduction to the molecular systematics of foraminifera. Micropaleontology 46(suppl 1):1–12

Pawlowski J, Burki F (2009) Untangling the phylogeny of amoeboid protists. J Eukaryot Microbiol 56:16–25

Pawlowski J, Lecroq B (2010) Short rDNA barcodes for species identification in foraminifera. J Eukaryot Microbiol 57:197–205

Pawlowski J, Bolivar I, Guiard-Maffia J, Gouy M (1994) Phylogenetic position of foraminifera inferred from LSU rRNA gene sequences. Mol Biol Evol 11:929–938

Pawlowski J, Bolivar I, Farhni J, Zaninetti L (1995) DNA analysis of *Ammonia beccarii* morphotypes: one or more species? Mar Micropaleontol 26:171–178

Pawlowski J, Holzman M, Fahrni J, Hallock P (2001) Molecular identification of algal endosymbionts in large miliolid foraminifers: 1. Chlorophytes. J Eukaryot Microbiol 48:362–367

Pawlowski J, Christen R, Lecroq B, Bachar D, Shahbazkia HR, Amaral-Zettler L, Guillou L (2011) Eukaryotic richness in the abyss: insights from pyrotag sequencing. PLoS ONE 6(4):e18169

Vlassov AV, Korba B, Farrar K, Mukerjee S, Seyhan AA, Ilves H, Kaspar RL, Leake D, Kazakov SA, Johnston BH (2007) shRNAs targeting hepatitis C: effects of sequence and structural features, and comparision with siRNA. Oligonucleotides 17:223–236

Chapter 7
FLEC-TEM: Using Microscopy to Correlate Ultrastructure with Life Position of Infaunal Foraminifera

Joan M. Bernhard and Elizabeth A. Richardson

Abstract To allow observation of cellular ultrastructure in foraminifers (or other meiofauna) preserved in their life position within sediments, modifications of the Fluorescently Labeled Embedded Core (FLEC) method are described. Such an approach will allow determinations of both fine-scale (sub-millimeter scale) distributional data and cell biological data on the same specimen.

Keywords Cell biology • Transmission electron microscopy • Epifluorescence microscopy • Confocal laser scanning microscopy • Sediment • Benthic • Protist

7.1 Introduction

The microbial world has many as yet unknown relationships. Some of these unknowns are likely to be fundamental in biogeochemical cycling, which is a critical underlying set of processes key to ecosystem functioning. Thus, it is important to discover the myriad of microbial relationships, both spatial and physiological, to further our understanding of biogeochemical cycling in the microbial world.

This contribution describes an extended application of the Fluorescently Labeled Embedded Core (FLEC) method, originally described by Bernhard and Bowser (1996), that fluorescently tags all metabolically active organisms and preserves them where they were living in sediments at the time of fixation. The application described here permits elucidation of the relationship between a specimen's submillimeter-scale

J.M. Bernhard (✉)
Geology and Geophysics Department, Woods Hole Oceanographic Institution, Woods Hole, MA 02543, USA
e-mail: jbernhard@whoi.edu

E.A. Richardson
Plant Biology Department, University of Georgia, Athens, GA 30602, USA

H. Kitazato and J.M. Bernhard (eds.), *Approaches to Study Living Foraminifera: Collection, Maintenance and Experimentation*, Environmental Science and Engineering, DOI 10.1007/978-4-431-54388-6_7, © Springer Japan 2014

distribution in sediments and its cell ultrastructure via Transmission Electron Microscopy (TEM). This extra step allows further documentation of cellular status and/or structure in specimens of established position at the sediment-water interface or within sediments at measurable depth. This information can assist, for example, with estimates of standing stock on the millimeter scale, identify microbe–microbe interactions, and survey for habitat-specific cellular adaptations.

This extended application is particularly valuable for investigations of benthic foraminifera given that it can, in certain instances, be problematic to distinguish living from dead individual foraminifers (Bernhard 1988, 2000).

7.2 Procedure

7.2.1 Core Collection and FLEC Embedding

Collect undisturbed small (~1.5 cm)-diameter syringe cores for analysis per the described FLEC procedure (Bernhard et al. 2003), ensuring that cores are maintained at or near in situ environmental conditions. Incubate each core in CellTracker™ Green CMFDA (Life technologies, Invitrogen) for 8–16 h depending on temperature. In general, colder conditions require longer incubation due to lower metabolic rates. After CellTracker Green incubation, move cores into a fume hood with adequate ventilation. Wearing personal protective equipment, infiltrate each core with TEM-grade glutaraldehyde to ensure a final concentration of ~3 %. Allow fixation for at least 24 h and then continue the FLEC preparation procedure through buffer rinses and serial ethanol dehydration steps as described originally (Bernhard et al. 2003). To promote best resin polymerization, omit the 1:1 acetone: uncatalyzed Spurrs' resin steps as described in Bernhard et al. (2003). Instead, transfer immediately from acetone to 100 % uncatalyzed Spurrs resin followed by at least two changes into fresh uncatalyzed resin and two catalyzed resin infiltrations. Be sure to use the "Long pot life" recipe for the Spurrs' embedding media (Electron Microscopy Sciences, Warrington, PA) to ensure a lower viscosity resin, which promotes infiltration through the entire core. Further details on methodology appear in Bernhard and Bowser (1996) and Bernhard et al. (2003).

Experience has shown that the FLEC method can substitute LR White resin for the originally used Spurrs' resin, although best FLEC results continue to be obtained with Spurrs'. If LR White is employed, polymerization with heat remains the best option (vs. microwave polymerization) in our trials. The impact, if any, of LR White substitution on cellular ultrastructure has not yet been determined.

7.2.2 FLEC Sectioning and Imaging

After the FLEC cores are polymerized, section each core in its desired orientation (e.g., sagittal or transverse) with either a hot-knife microtome (McGee-Russell

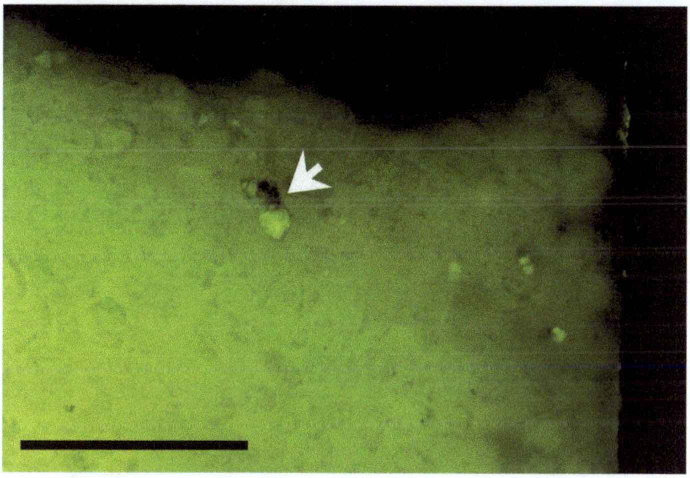

Fig. 7.1 Epifluorescence micrograph showing a portion of a FLEC core section from a Type 1 stromatolite. Visible is the sediment-water interface and one edge of the core. The largest brightly fluorescent object is a calcareous foraminifer (*arrow*) that was selected for TEM analysis. Scale = 1 mm

et al. 1990) or with a low-speed rock saw with the thinnest available diamond blade (e.g., 1 mm, Isomet). Thicker blades necessitate more sample loss, which can be an issue in some applications. The method of choice for sectioning largely depends on grain size, with coarsely grained sediments requiring the rock-saw approach.

Using epifluorescence microscopy (480-nm excitation, 520-nm emission), scan both sides of each FLEC core section and document the position of all foraminifera (and/or other microbes of interest; Fig. 7.1). Such data can be used to compile fine-scale distributions (Bernhard et al. 2003). Next, image selected specimens with higher magnification using Laser Scanning Confocal Microscopy (LSCM; Fig. 7.2), which can reveal many more details than observable with epifluorescence microscopy (Bernhard et al. 2003, 2013; First and Hollibaugh 2010). Select specimens that are particularly of interest, due perhaps to a particular position, morphology, or interaction, for investigation of cellular ultrastructure.

7.2.3 Excision and Preparation for TEM Sectioning

Using the previously obtained epifluorescence micrograph(s) and epifluorescence microscopy with the same filter set, find the specimen of interest in the appropriate FLEC section. Keeping in mind that parts of the FLEC section likely will fracture during specimen removal, ensure that all desired imaging of the section, particular specimen, and specimen's surroundings are obtained prior to specimen excision. Excise the targeted specimen with a very sharp razor blade, keeping ~1 mm margins on each side of the individual.

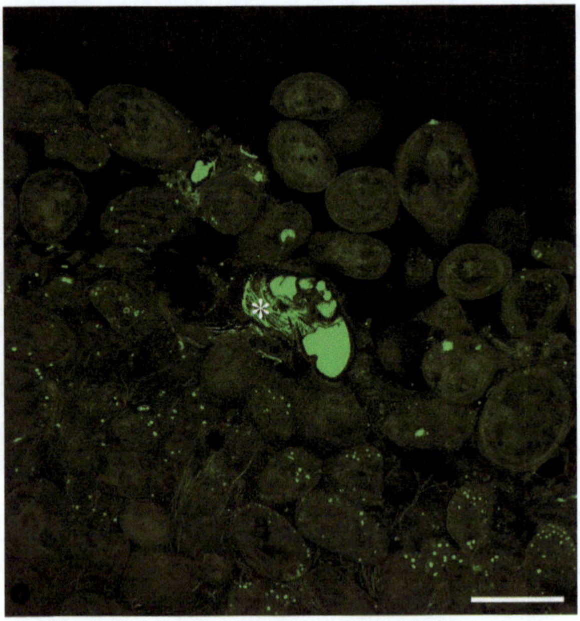

Fig. 7.2 Laser scanning confocal micrograph of the foraminifer shown in Fig. 7.1. The foraminifer had frothy cytoplasm in the youngest chamber (*). Also note the other visible microbes. Scale = 200 μm

Again using the epifluorescence microscope, trim the excised material into a block for microtomy (Reid 1975) using another sharp razor blade. Typically, the block is a trapezoidal pyramid, trimmed so that the edges come close but do not remove any portion of the foraminiferan (Fig. 7.3). Next, using an ultramicrotome and diamond knife, serial section the trimmed specimen into 80-nm or thicker sections. On slot or mesh Formvar-coated grids, mount sections and allow to air dry. The number of sections that will fit on each grid depends on the surface area of the block. Stain the grids with a saturated solution of uranyl acetate (50 % in ethanol) and lead stain (Hanaichi et al. 1986) and allow to air dry.

7.2.4 Imaging Cellular Ultrastructure with TEM

View the stained sections with a TEM, as appropriate (e.g., 80 KV). By scanning the section(s) at low magnification, if desired, one can determine the location of the current field of view with respect to the orientation of the foraminifer in the earlier obtained epifluorescence micrographs or LSCM images. Orientation is easier to establish with slot grids due to the increased unobstructed surface area, but the Formvar on slot grids typically fails more often than that on mesh grids. Image cellular structures with TEM as desired (Fig. 7.4).

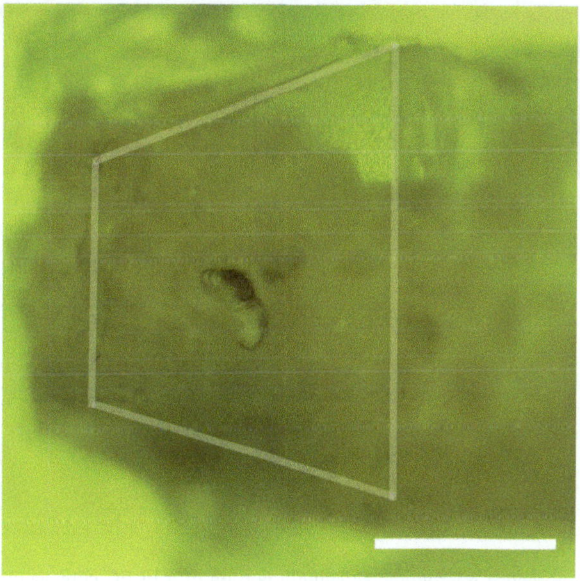

Fig. 7.3 Transmitted light micrograph showing the foraminifer from Figs. 7.1 and 7.2 after excision from the FLEC section. Sedimentary grains (ooids) remain in place around the foraminifer. Note the trapezoidal shape of the block, outlined by the *gray lines*. Scale = 500 μm

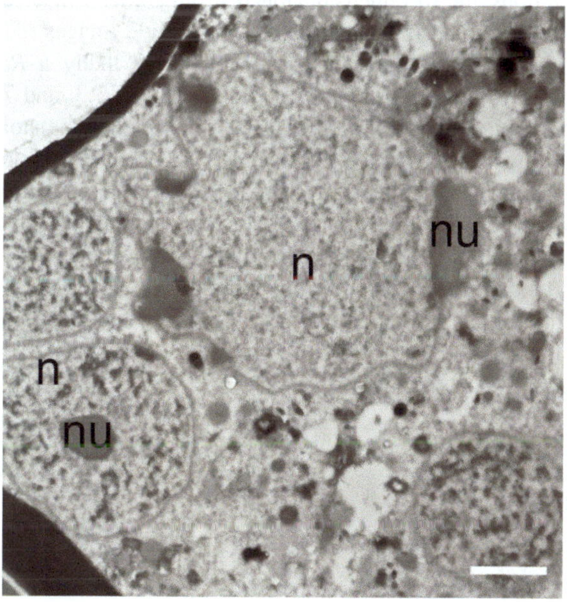

Fig. 7.4 Transmission electron micrograph showing a portion of the foraminifer from Figs. 7.1, 7.2, and 7.3. Visible are two nuclei (n) each with a nucleolus (nu); two other nuclei are also likely. Scale = 2 μm

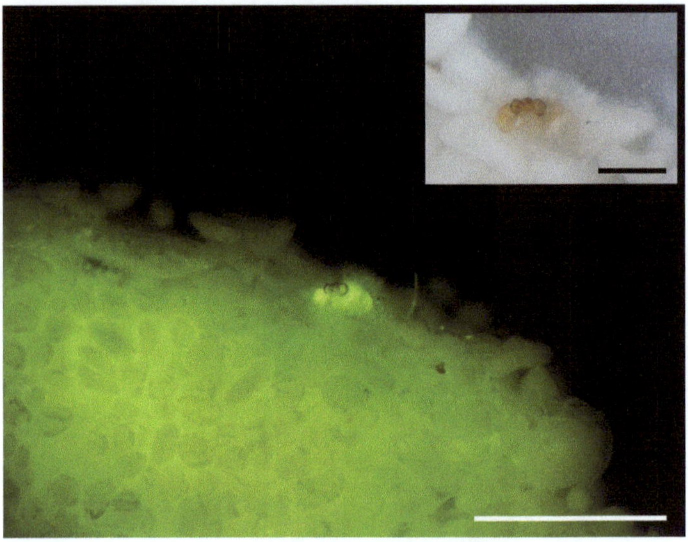

Fig. 7.5 Epifluorescence micrograph showing a portion of a FLEC core section from a Type 2 stromatolite. Visible is the sediment-water interface and the foraminifer, which is the brightly fluorescent object that was selected for TEM analysis. Scale = 1 mm. *Inset*: reflected light micrograph of the same specimen. Inset scale = 250 μm

In our illustrations, the specimens inhabited stromatolites collected off Highborne Cay, Bahamas. For demonstration purposes, we include images of two calcareous specimens, probably of the same species, which was likely a *Rosalina* species (Bernhard et al. 2013). The first specimen (Figs. 7.1, 7.2, 7.3, and 7.4), which was <0.5 mm below the sediment-water interface of a Type 1 stromatolite (Reid et al. 2000), was multinucleate, as evidenced by Fig. 7.4. The second specimen, which inhabited the sediment-water interface of a Type 2 stromatolite (Reid et al. 2000) (Fig. 7.5), had what appear to be copious reticulopods emanating from its umbilical side (Fig. 7.6) and multiple nuclei (Fig. 7.7). Mitochondria appear to be lightly stained, but cristae can be discerned (Fig. 7.8). Food vacuoles were also apparent (Fig. 7.8). For more information on the foraminiferal assemblages and distributions as well as protist populations in these Bahamian microbial mats, see Bernhard et al. (2013) and Edgcomb et al. (2013), respectively.

7.3 Discussion

7.3.1 Applications

This modification of the FLEC method (Bernhard et al. 2003; First and Hollibaugh 2010; Edgcomb and Bernhard 2013) pairs cytological and ecological data on any given specimen. For example, in chemocline habitats where short vertical distances

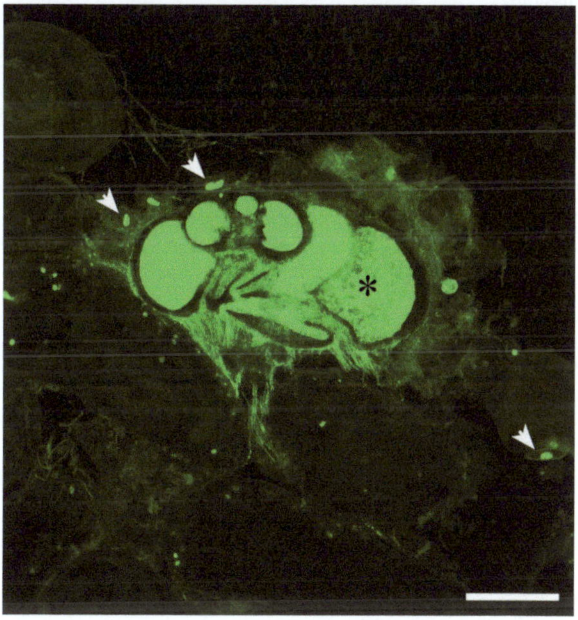

Fig. 7.6 Laser scanning confocal micrograph of the foraminifer shown in Fig. 7.5. The foraminifer had frothy cytoplasm in the youngest chamber (*). Also note the flagellates (*arrowheads*). Scale = 100 μm

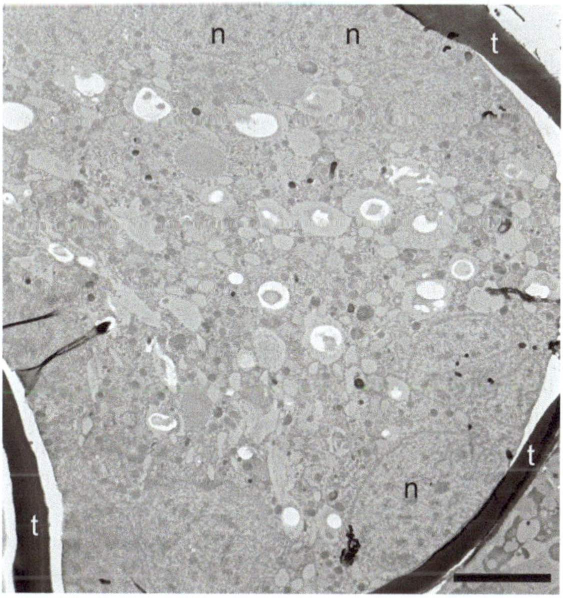

Fig. 7.7 Transmission electron micrograph showing a portion of the foraminifer from Figs. 7.5 and 7.6 after excision from the FLEC section. Note the multiple nuclei (n) and remnant calcareous test (t). Scale = 5 μm

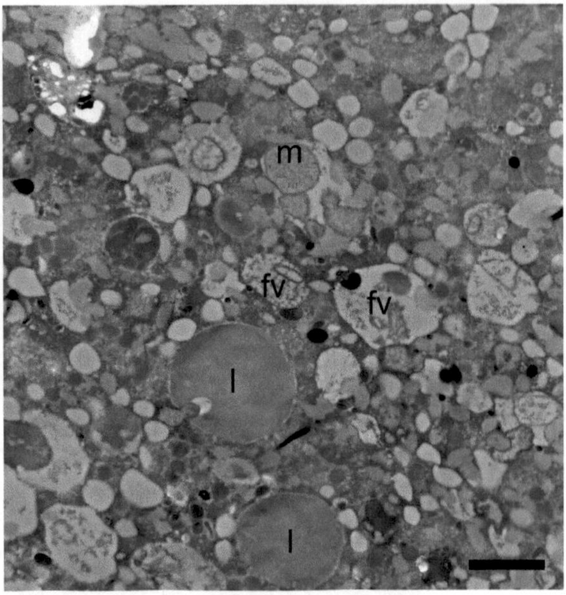

Fig. 7.8 Transmission electron micrograph showing a portion of the foraminifer from Figs. 7.5, 7.6, and 7.7. Note the cristae in the mitochondrion (m). Also visible are food vacuoles (fv) and lipid (l). Scale = 2 μm

have a significant impact on the geochemistry of pore waters (e.g., McCorkle et al. 1990; Visscher et al. 2000; Bernhard et al. 2003), it may be important to establish if the foraminifera at a given sediment depth are indeed living. Ultrastructural analysis provides a tool to resolve this question. As has been shown at cold methane seeps, not all Rose-Bengal stained foraminifera were living at the time of collection as evidenced by ultrastructural analysis (Bernhard et al. 2010). Thus, this novel FLEC-TEM approach will help resolve foraminiferal ecology and cell biology in so-called "extreme" habitats such as seeps and sediment-covered hydrothermal vents. Furthermore, the approach can be used in any soft substrate where insights into fine-scale benthic ecology and cytology are warranted, such as studying the impact of catastrophic events such as the Deepwater Horizon oil spill.

The application is particularly useful for investigations regarding microhabitat-specific cellular adaptations. Organelle attributes can be examined in specimens from particular horizons. For example, peroxisome-endoplasmic reticulum complexes have been noted in foraminifera from sulfidic sediments (e.g., Bernhard and Bowser 2008) but fine-scale distributions of specimens exhibiting these organelles with respect to the sediment-water interface or chemocline have not been conducted. Another habitat-specific adaptation that FLEC-TEM can help elucidate is putative symbiosis. Endobionts or ectobionts might be expected in foraminiferan hosts inhabiting sulfidic and/or anoxic to hypoxic (oxygen depleted) horizons (e.g., Bernhard et al. 2006). The FLEC-TEM approach can be used in combination with microelectrode profiling to establish if putative symbionts are supported in any particular species from a given geochemical microhabitat. Thus, the FLEC-TEM approach assists identifications of microbe-microbe associations and interactions.

Also, using this approach it may be possible to identify reproductive events early in the process when gametes begin to form (Goldstein et al. 2010) or prior to dispersal of propagules from the parental test (Goldstein and Alve 2011). Such insights will aid in determining if reproduction is linked to specific depth horizons or other spatial characteristics.

This FLEC-TEM approach can also help identify diet and feeding location. For example, in FLEC images where diatoms are gathered adjacent to a foraminiferan's aperture (e.g., Bernhard et al. 2013 Fig. 1D), FLEC-TEM could provide further evidence of active feeding by revealing food-vacuole contents.

7.3.2 Caveats and Notes Regarding Methodology

Typically, cellular ultrastructural investigations use osmium exposure post fixation prior to embedding to enhance membrane visibility. In the case of FLEC, osmication prior to embedding is not possible because it quenches fluorescence. Membrane contrast in the TEM appeared adequate using the recommended sectioning and staining protocol. Another disadvantage to osmium post-fixation includes generation of toxic waste.

Correlative epifluorescence/TEM in sectioned specimens was advanced by Bowser and Rieder in the 1980s. Rieder and Bowser (1987) presents a review of additional methods of correlative fluorescence and TEM.

Because FLEC cores are not decalcified at any point during the preparation procedure and because other (e.g., siliceous) minerals are not removed prior to microtomy, sectioning of blocks containing the foraminifera and surrounding sediment grains could dull or damage the diamond knife. The cost of sharpening a diamond knife is significant and such an expense should be considered before undertaking a FLEC-TEM project. Thicker (e.g., 150 nm) sections cut with less expensive "histo" grade diamond knives can reveal useful ultrastructural information using conventional (80–120 kV) TEM, and should be explored as an alternative measure.

It is also important to note that the incubation period for the CellTracker Green CMFDA can be extensive (many hours, depending upon temperature). Thus, this approach might be inappropriate in cases where the environmental conditions cannot be maintained at or near in situ conditions during the entire incubation period. Conditions that differ from those in situ could cause artifacts in distributions or physiology, both of which would be manifest in the FLEC-TEM analyses. For example, specimens may migrate or incidentally-captured metazoans can disturb sediment and/or foraminifera. Ideally such factors would be monitored during the incubation period.

Acknowledgements Support for this research was provided by the US NSF grant OCE-0926421 to J.M.B. and V.P. Edgcomb. We thank Sam Bowser for insightful discussions and helpful comments on an earlier manuscript version. J.M.B. also thanks JAMSTEC and Tohoku University for financial support to allow her contributions to the Field Workshop on Living Foraminifera in Japan in 2012.

References

Bernhard JM (1988) Postmortem vital staining in benthic foraminifera: duration and importance in population and distribution studies. J Foramin Res 18:143–146

Bernhard JM (2000) Distinguishing live from dead foraminifera: methods review and proper applications. Micropaleontology 46:38–46

Bernhard JM, Bowser SS (1996) Novel epifluorescence microscopy method to determine life position of foraminifera in sediments. J Micropalaentol 15:68

Bernhard JM, Bowser SS (2008) Peroxisome proliferation in foraminifera inhabiting the chemocline: an adaptation to reactive oxygen species exposure? J Eukaryot Microbiol 55:135–144. doi:10.1111/j.1550-7408.2008.00318.x

Bernhard JM, Visscher PT, Bowser SS (2003) Submillimeter life positions of bacteria, protists, and metazoans in laminated sediments of the Santa Barbara Basin. Limnol Oceanogr 48:813–828

Bernhard JM, Habura A, Bowser SS (2006) An endobiont-bearing allogromiid from the Santa Barbara Basin: implications for the early diversification of foraminifera. J Geophys Res Biogeosci 111:G03002. doi:10.1029/2005gj000158

Bernhard JM, Martin JB, Rathburn AE (2010) Combined carbonate carbon isotopic and cellular ultrastructural studies of individual benthic foraminifera: 2. Toward an understanding of apparent disequilibrium in hydrocarbon seeps. Paleoceanography 25:Pa4206. doi:10.1029/2010pa001930

Bernhard JM, Edgcomb VP, Visscher PT, Mcintyre-Wressnig A, Summons R, Bouxsein ML, Louis L, Jeglinski M (2013) Insights into foraminiferal influences on microfabrics of microbialites at Highborne Cay, Bahamas. Proc Natl Acad Sci USA. doi:10.1073/pnas.1221721110

Edgcomb VP, Bernhard JM (2013) Heterotrophic protists in hypersaline microbial mats and deep hypersaline Basin water columns. Life 3:346–362. doi:10.3390/life3020346

Edgcomb VP, Bernhard JM, Summons RE, Orsi W, Beaudoin D, Visscher PT (2013) Active eukaryotes in microbialites from Highborne Cay, Bahamas and Hamelin Pool (Shark Bay), Australia. ISME J. doi:10.1038/ismej.2013.130

First MR, Hollibaugh JT (2010) Diel depth distributions of microbenthos in tidal creek sediments: high resolution mapping in fluorescently labeled embedded cores. Hydrobiologia 655:149–158. doi:10.1007/s10750-010-0417-2

Goldstein ST, Alve E (2011) Experimental assembly of foraminiferal communities from coastal propagule banks. Mar Ecol Prog Ser 437:1–11. doi:10.3354/meps09296

Goldstein ST, Habura A, Richardson EA, Bowser SS (2010) *Xiphophaga minuta* and *X. allominuta*, nov. gen., nov. spp., new monothalamid foraminifera from coastal Georgia (USA): cryptic species, gametogenesis, and an unusual form of chloroplast sequestration. J Foramin Res 40:3–15

Hanaichi T, Sato T, Iwamoto T, Malavasiyamashiro J, Hoshino M, Mizuno N (1986) A stable lead by modification of Sato method. J Electron Microsc 35:304–306

McCorkle DC, Keigwin LD, Corliss BH, Emerson SR (1990) The influence of microhabitats on the carbon isotopic composition of deep-sea benthic foraminifera. Paleoceanography 5(2):161–185

McGee-Russell SM, Debruijn WC, Gosztonyi G (1990) Hot knife microtomy for large area sectioning and combined light and electron microscopy in neuroanatomy and neuropathology. J Neurocytol 19:655–661. doi:10.1007/bf01188034

Reid N (1975) Trimming the specimen block. In: Glauert AM (ed) Practical Methods in Electron Microscopy. North-Holland Publishing, Amsterdam

Reid RP, Visscher PT, Decho AW, Stolz JF, Bebout BM, Dupraz C, Macintyre LG, Paerl HW, Pinckney JL, Prufert-Bebout L, Steppe TF, DesMarais DJ (2000) The role of microbes in accretion, lamination and early lithification of modern marine stromatolites. Nature 406:989–992

Rieder CL, Bowser SS (1987) Correlative light and electron microscopy on the same epoxy section. In: Hayat MA (ed) Correlative Microscopy in Biology: Instrumentation and Methods. Academic, Orlando

Visscher PT, Reid RP, Bebout BM (2000) Microscale observations of sulfate reduction: correlation of microbial activity with lithified micritic laminae in modern marine stromatolites. Geology 28:919–922. doi:10.1130/0091-7613(2000)28<919:moosrc>2.0.co;2

Chapter 8
Response of Shallow Water Benthic Foraminifera to a ^{13}C-Labeled Food Pulse in the Laboratory

V.N. Linshy, Rajiv Nigam, and Petra Heinz

Abstract In recent years, stable isotope techniques have been applied successfully directly on the sea floor to follow the fate of carbon in the marine carbon cycle. Feeding experiments using ^{13}C-labeled food enable the tracking of ingested tracer material in the cytoplasm of foraminifera to study phytodetrital carbon uptake in situ. Here we introduce first results of a ^{13}C-labeled food pulse in the laboratory. Benthic foraminiferal species *Ammonia tepida* and *Bolivina variabilis* were studied to analyze their response. Marked green algae *Dunaliella tertiolecta* were added at the beginning of the experiment and the foraminiferal reaction was followed for 42 days. In order to assess the uptake of labeled material with time, foraminiferal specimens were subsampled at 2, 7, 21, and 42 days and analyzed for their carbon signals. Culture conditions (T, salinity, O_2) were monitored and kept stable during time. Both species showed fast and strong ingestion of the added food material after 2 days, followed by a decreasing isotopic signal in the biomass with time. Additionally, the isotopic signals in the overlying water of the experimental dishes and in the foraminiferal tests were measured, but these values were difficult to interpret.

8.1 Introduction

Benthic foraminifera form a large part of the biomass of the marine benthic community (Heip et al. 2001) and are important proxies for paleoenvironmental studies owing to their abundance, sensitivity to environmental changes and extensive

V.N. Linshy (✉) • R. Nigam
Micropaleontology Laboratory, National Institute of Oceanography, CSIR, Dona Paula, Goa 403 004, India
e-mail: linshy@gmail.com

P. Heinz
Department of Palaeontology, University of Vienna, Vienna, Austria

H. Kitazato and J.M. Bernhard (eds.), *Approaches to Study Living Foraminifera: Collection, Maintenance and Experimentation*, Environmental Science and Engineering, DOI 10.1007/978-4-431-54388-6_8, fossilization potential. At times they can comprise more than 50 % of the benthic biomass (Snider et al. 1984; Gooday et al. 1992) and therefore play a significant role in the carbon budget of the oceans. Oceans hold about 60 times more carbon than in the atmosphere and therefore a small change in the chemistry of oceans can produce a marked change in atmospheric reservoir (Berger et al. 1989). Ocean productivity helps to control the partitioning of carbon between the ocean reservoir and the atmospheric reservoir (Berger et al. 1989). Therefore, ocean productivity changes play an important role in providing feedback to climatic changes. In the oceans, primary production is carried out by marine microscopic algae, which use sunlight as an energy source by converting inorganic carbon to organic carbon. The organic carbon that reaches the sea floor either accumulates in the sediment or is decomposed by benthic organisms. Hence it is important to understand the activities and population dynamic of the benthos, including meiofauna, to understand the cycling of organic carbon on a global scale (Shimanaga et al. 2000). Meiofauna, which mainly consists of small sized metazoa and foraminifera, make a significant part of sea floor biomass and play an important role in the benthic metabolism (Moodley et al. 2000).

Supply of organic flux to the ocean sediments and foraminiferal abundance are positively correlated in most parts of the world, e.g., Eastern equatorial Pacific (Pedersen et al. 1988), Western equatorial Pacific (Herguera 1992), Adriatic Sea (Jorissen et al. 1992), NE Atlantic (Gooday 1993), South China and Sulu seas (Miao and Thunell 1993; Rathburn and Corliss 1994), East China Sea (Wahyudi and Minagawa 1997), Arctic, Atlantic, Pacific and Indian Oceans (Altenbach and Struck 2001). The surface ocean productivity is reported to control the organic carbon flux to the ocean floor, which in turn governs the benthic foraminiferal assemblages of given areas (Corliss and Emerson 1990; Gooday and Turley 1990; Ohga and Kitazato 1997; Gooday and Rathburn 1999; Altenbach et al. 1999; Kitazato et al. 2000; Fontanier et al. 2002, 2003).

Infrequent or irregular deposition of organic carbon pulses into sea floor influence the activities and feeding strategies of benthic organisms (Gooday and Turley 1990; Jumars et al. 1990; Pfannkuche 1993; Smith et al. 1994). Phytodetritus and its more labile components are among the main food sources for benthic organisms (Gooday and Lambshead 1989; Altenbach 1992). It forms a suitable food source for many foraminifera and affects their faunal abundance. However quantitative information on the ingestion rate of organic carbon by benthic foraminifera has remained scarce and feeding ecologies in relation to phytodetritus processing are not well understood (Nomaki et al. 2006). It is necessary to clarify the fate of the deposited phytodetritus by biological as well as geochemical processes to understand the carbon budget of the ocean floor.

Experimental approaches are helpful to understand short-term responses of benthic foraminifera to phytodetritus deposition. Experiments where phytodetritus is artificially introduced into the system can enable observations on the feeding behavior of foraminifera on a time scale of days to weeks. Stable isotopes are useful tracers for following the fate of labeled material over short- to long-time scales and for examining quantitatively the ingestion rates of benthic organisms (Blair et al. 1996; Levin et al. 1999; Middelburg et al. 2000). In in situ feeding experiments ^{13}C-labeled phytodetritus have become an effective tool to study the phytodetrital carbon uptake by benthic foraminifera. Ingested tracer material changes the isotopic composition of the foraminiferal cytoplasm and analyses of the biomass allow quantitative statements about ingestion rates of organic carbon by the foraminifera. Successful in situ feeding experiments were reported from an intertidal flat in the southern North Sea (Moodley et al. 2000), a fjord in Norway (Witte et al. 2003a; Sweetman et al. 2009), the bathyal Sagami Bay in the Northwest Pacific (Nomaki et al. 2005a, 2006, 2009), the abyssal North Atlantic (Moodley et al. 2002; Witte et al. 2003b), Station M in the eastern North Pacific (Enge et al. 2011) and the OMZ in the Arabian Sea (Woulds et al. 2007; Andersson et al. 2008). These investigations demonstrate the active and important role of benthic foraminifera in cycling organic material, although the range of ingestion rates can strongly vary between different habitats and foraminiferal assemblages.

In addition to in situ investigations, laboratory feeding experiments can help to better understand the foraminiferal response to a given food pulse. Laboratory experiments allow modification of one single environmental parameter (in this case, food availability), while other factors (like oxygen supply) can be monitored and kept constant (Heinz et al. 2001, 2002). This ability to control the environment is advantageous especially for long-term studies. Moreover, the experimental structure enables analysis of the reaction of an isolated group or even single species that were obtained from the original sediment and transferred into the experimental setup. Despite these advantages, as far as we know, no laboratory study using isotopical labeled food material has been reported in the literature for benthic foraminifera to date.

In the present investigation, we tested the response of two shallow-water benthic foraminifera to a ^{13}C-labeled food pulse under experimental lab conditions. At the beginning of the experiment, ^{13}C-labeled green algae (*Dunaliella tertiolecta*) were added into the setup. Uptake rates of *Ammonia tepida* and *Bolivina variabilis* were determined in a time series (2–42 days) to analyze and compare the response rates over time for both species under study.

8.2 Material and Methods

8.2.1 Sample Collection and Preparation

Ammonia tepida (Cushman 1926) was collected from an intertidal mud flat in the Bay of Aiguillon, near La Rochelle, France (Fig. 8.1). The sediment samples were carefully scooped into sample containers and were briefly washed over 63 μm sieve with seawater at the collection site itself in order to remove part of the muddy fraction and to concentrate the foraminiferal samples. Thereafter, the samples were properly

Fig. 8.1 Various steps in the initial treatment of sediment samples in the field: (**a**) Sediment samples being scooped in to sampling box. (**b**) Closer look of the collected sample. (**c**) Preliminary sieving of the samples in the field. (**d**) Sieved sampled getting transferred to bottles along with seawater

washed over 63 μm sieve with seawater collected from the sampling site later in the laboratory of the University of Angers. The residues were transferred to pre-labeled plastic containers with adequate amount of seawater and transported to the Micropaleontology laboratory in Tuebingen, Germany. Washing the samples before transportation removes most of the fine material containing high amounts of organic carbon and helps to avoid oxygen deficiency because of bacterial degradation. After samples arrived in the Tuebingen lab, living *A. tepida* individuals were picked out of the sediment, cleaned from sediment residues and transferred into the experimental dishes (Fig. 8.2).

Bolivina variabilis (Williamson 1858) was collected from a culture that has been maintained in the Tuebingen lab since 1988. Originally, the sediments for this culture were collected from the coastal water off Kenyan. *Bolovna variabilis* species live in high densities on algal mats that were formed during the years in this culture (Fig. 8.3a, b). Individuals were picked from the algae mats, cleaned from algal residues and transferred into the experimental dishes (Fig. 8.3c, d).

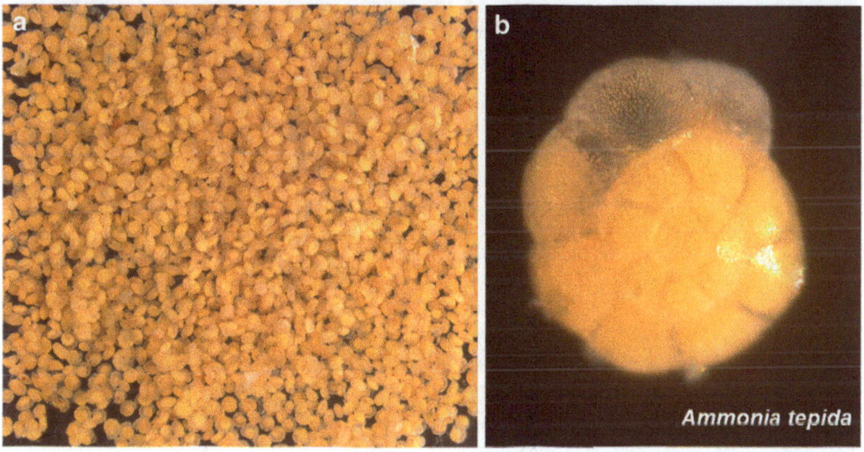

Fig. 8.2 (**a**) Living *Ammonia tepida* individuals picked out of the sediment. (**b**) Isolated specimen of *A. tepida*

8.2.2 Production of ^{13}C-Labeled Algae

The unicellular green algae *D. tertiolecta*, was used as ^{13}C-labeled food marker. The algae were cultured at 20 °C in an incubator with a 12 h light–12 h dark illumination cycle. *Dunaliella tertiolecta* was grown in prefiltered and sterilized natural seawater amended with f/2 medium (after Giullard and Ryther 1962). For labeling, 0.1 mol l^{-1} ^{13}C-bicarbonate (99 atom% ^{13}C-enriched $NaHCO_3$, Chemotrade Leipzig) was added. The algal material was harvested by centrifugation (4,000 g, 10 min) and washed three times with sterile-filtered (0.2 μm) seawater to remove excess $NaH^{13}CO_3$ and any dissolved organic ^{13}C exuded by the green algae. Algae were frozen in liquid nitrogen afterwards to avoid the lysis of cells and stored at −80 °C until freeze-drying. The final ^{13}C concentration in the algae was 75 atom% ^{13}C.

8.2.3 Experiment Set-Up

Two time series were incubated for these investigations: one for *A. tepida* and one for *B. variabilis*. For each time series, four time intervals (2, 7, 21, 42 days) were prepared in replicate totaling 16 experimental dishes (Fig. 8.4). Each dish contained 500 specimens and 100 ml of seawater. All specimens were active and tests were filled with intact protoplasm, looking very healthy. At the beginning of the experiment (t=0), labeled algal food was introduced (614 $mgCm^{-2}$) in each dish. All dishes were incubated at 20 °C in an incubator with a 12 h light–12 h dark

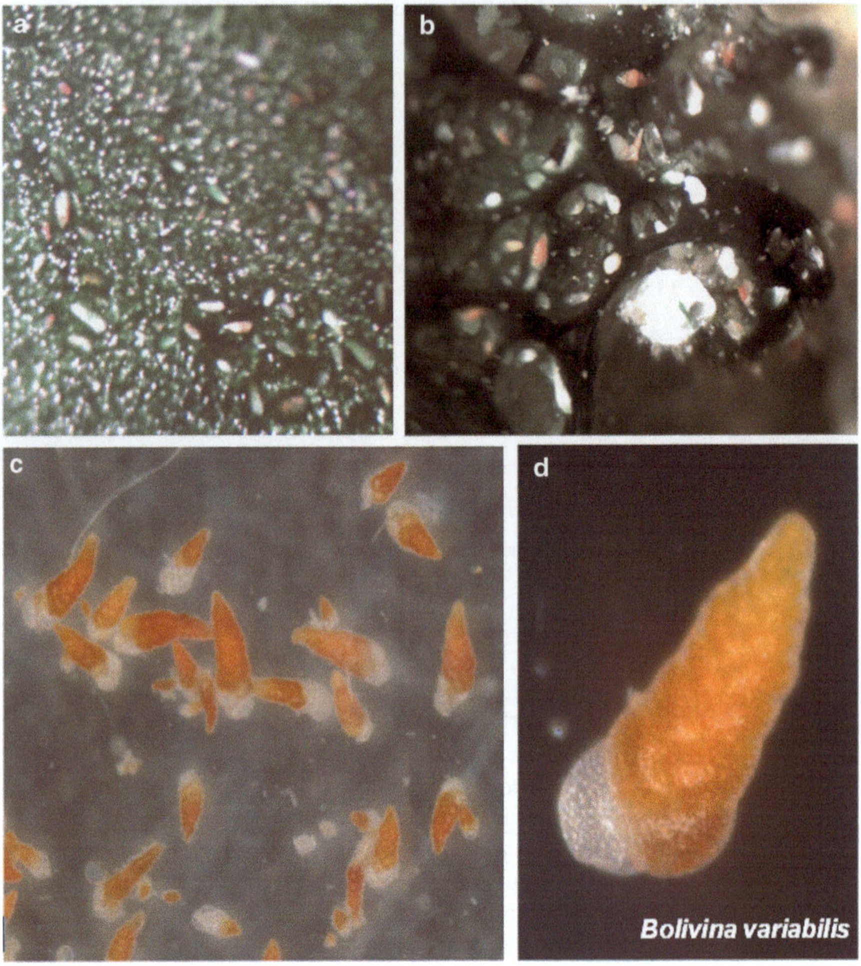

Fig. 8.3 (**a**) Living *Bolivina variabilis* in the laboratory culture. (**b**) Closer look to the algal mat containing *B. variabilis*. (**c**) Individuals picked out of the algal mat (**d**) Isolated specimen of *B. variabilis*

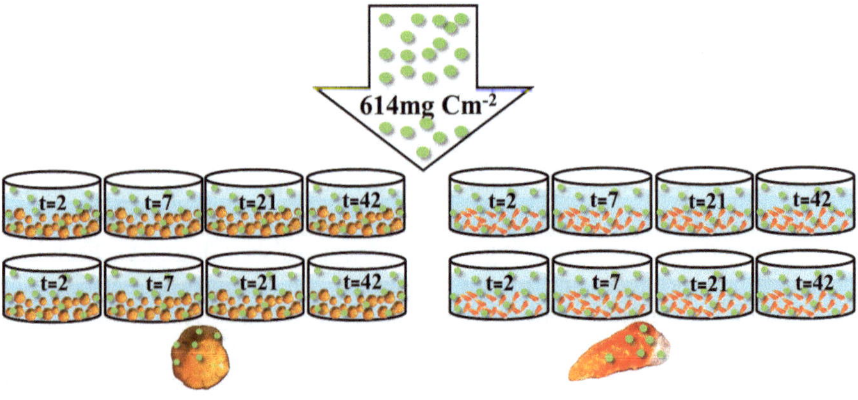

Fig. 8.4 Experimental setup of the two time series. Each dish contains 500 living specimens of either *A. tepida* or *B. variabilis*

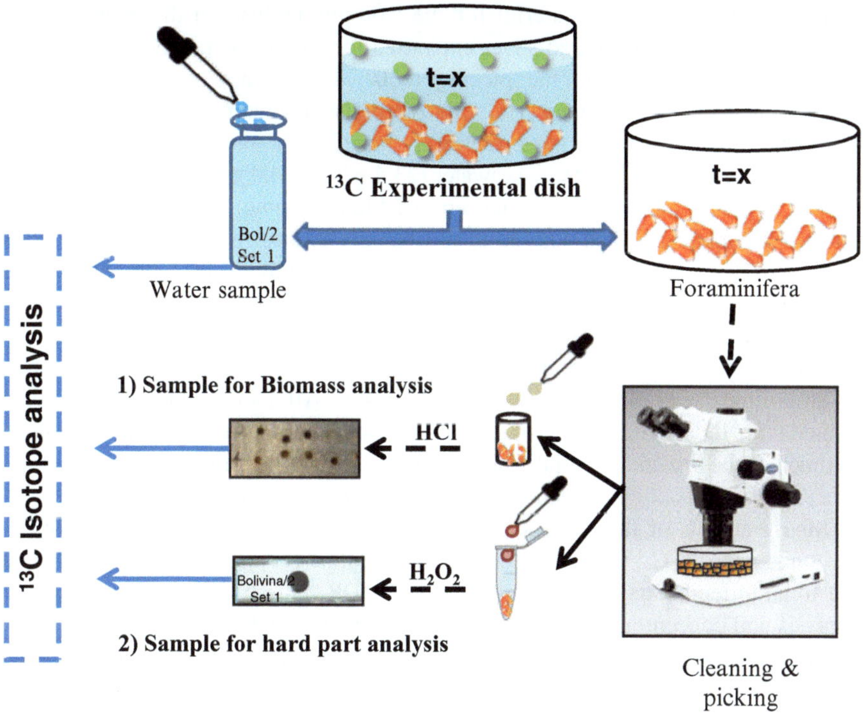

Fig. 8.5 Overview of sample preparation after incubation

illumination cycle. Salinity was measured regularly with a conductivity meter LF 340 (WTW, Germany) and was stable of the order of 34 ± 1 ‰. Dissolved oxygen was measured regularly with an oxygen microelectrode and a picoamperemeter (MasCom, Bremen, Germany). Linear calibration of the electrode signal was obtained with an oxymeter OXI 340/A (WTW, Germany) with air-saturated seawater and oxygen-free seawater aerated with nitrogen. The oxygen content was stable of the order of 7 ± 1 ml/l throughout the experiment.

8.2.4 *Sample Preparation and Measurement*

After incubation, the dishes were taken out of the incubator and samples were prepared for various analyses. For determination of the stable carbon isotope ratio in the overlying water, water was transferred into 50-ml glass vials, poisoned with some drops of a saturated solution of $HgCl_2$, tightly sealed free from air bubbles and kept cool until the measurements. The foraminifera were carefully removed from the experimental dishes. 350 out of the total 500 specimens in each set were kept for the biomass analysis and the remaining 150 specimens were used for studying the isotopic signal in the foraminiferal tests (Fig. 8.5).

Before processing the material for the isotopic analysis in the foraminiferal biomass, all the glassware and silver cups used for various preparations were combusted (450 °C, 5 h) and the picking tools were cleaned with a mixture of Dichloromethane and Methane (1:1, v:v) to remove adhered organics. Foraminifera were carefully cleaned of organic matter by brushing the outer surface of the test, washed twice in filtered artificial seawater (23.4 g NaCl, 4.0 g $MgSO_4 \times 7\ H_2O$, 0.8 g KCl, 0.26 g $CaCl_2$ and distilled water up to 1:l final volume) and finally transferred into silver cups (each filled with 30 µl of filtered seawater). Subsequently, the filled cups were dried at 50 °C for several hours before adding 20 µl hydrochloric acid (10 %) to remove all calcium carbonate in the samples. Heating and acidification were repeated once more before keeping samples at 50 °C for 3 days until completely dry.

For the analysis of foraminiferal tests, the specimens were initially cleaned two to three times with a brush in artificial seawater. Afterwards, the samples were treated with 10 % hydrogen peroxide and ultrasonicated for few seconds to remove as much of the organic material inside the tests as possible. In spite of this cleaning process, remnants of cytoplasm were recognizable in several tests.

Measurements of the total organic carbon (TOC) content and the ratio of the carbon isotopes ($^{13}C/^{12}C$) in the water samples, in the foraminiferal cytoplasm, and the foraminiferal tests were performed by PD DR. Ulrich Struck. Stable isotope analysis and concentration measurements of carbon were be performed simultaneously with a THERMO/Finnigan MAT V isotope ratio mass spectrometer, coupled to a THERMO Flash EA 1112 elemental analyzer via a THERMO/Finnigan Conflo III- interface in the stable isotope laboratory of the Museum für Naturkunde, Berlin. Stable isotope ratios were expressed in the conventional delta notation ($\delta^{13}C$) relative to VPDB (Vienna PeeDee Belemnite standard). Standard deviation for repeated measurements of lab standard material (peptone) was generally better than 0.15 per mill (‰) for carbon. Standard deviations of concentration measurements of replicates of lab standard was <3 % of the concentration analyzed.

8.2.5 Calculation of Carbon Uptake

The carbon isotope ratio ($^{13}C/^{12}C$) of the foraminiferal cytoplasm was measured against the International Vienna Pee Dee Belemnite standard (VPDB) and is expressed as the difference between the sample and the standard in the δ-notation: $\delta^{13}C\ [‰] = [(^{13}C/^{12}C_{sample})/(^{13}C/^{12}C_{VPDB}) - 1] \times 10^3$. The specific uptake of labeled ^{13}C by foraminifera or algae was calculated as excess (above background) and is expressed in the Δ-notation: $\Delta\delta^{13}C\ [‰] = \delta^{13}C_{sample} - \delta^{13}C_{background}$. The amount of carbon uptake was calculated as the product of the total sample excess and the carbon content of the cytoplasm in the sample: ($\Delta\delta^{13}C$ foraminifera – $\Delta\delta^{13}C$ algae) * Biomass. According to the calculations of Nomaki et al. (2005a, 2006), we can determine the fraction f of carbon originated from ^{13}C-labeled food incubated for a specific time interval: (atom%^{13}C foram – atom%^{13}C foram background)/(atom%^{13}C alga – atom%^{13C} algae background).

8.3 Results

Table 8.1 shows the analyzed isotopic measurements ($\delta^{13}C$) and total organic carbon content (TOC) for *A. tepida* and *B. variabilis* incubated for different time intervals as an average of both replicates. For each series and each time point, 350 individuals were measured. The isotopic signatures of benthic foraminifera just before addition of labeled food (t=0) exhibit background $\delta^{13}C$ values of −16.39‰ for *A. tepida* and −17.33‰ for *B. variabilis*. In both time series, high $^{13}C/^{12}C$-ratios in the cytoplasm of living foraminifera in incubated sediment compared to the background can be recognized. For *A. tepida*, highest carbon excess was recorded after 2 days, for *B. variabilis* after 7 days. Highly positive $\Delta\delta^{13}C$ values were interpreted as evidence for the ingestion of ^{13}C-labeled algae material. Values of the TOC content in the biomass varied between species and incubation times. *Ammonia tepida* generally showed higher TOC values than *B. variabilis*, with a maximum at the beginning of the time series and decreasing TOC contents with increasing incubation time. *Bolivina variabilis* recorded variable TOC values, with maxima after 2 and 21 days of incubation. Labeled carbon uptake calculated from ^{13}C excess and TOC content was highest after 2 days of incubation in both time series, although *B. variabilis* shows high variations between both replicates and therefore high standard deviation at specific time intervals (Fig. 8.6). In average, the uptake level was slightly higher for *B. variabilis*, and even after 42 days elevated carbon uptake and incorporation of algae material was recognized here. The calculated fraction of carbon originated from ^{13}C-labeled food recorded the same trend in *A. tepida* compared to the uptake, with a maximum after 2 days and decreasing numbers with increasing incubation time, while *B. variabilis* shows a maximum after 7 days, comparable to the analyzed isotopic measurements ($\delta^{13}C$) (Fig. 8.7).

The isotopic signal in the overlaying water of the experimental dishes was measured (Fig. 8.8). In the *A. tepida*-series, highest $\delta^{13}C$-values in the water caused by decomposition processes of the ^{13}C-labeled algae were detected after 7 days. Afterwards, the signal is strongly weakened. For *B. variabilis*, the water signal differs. Constant high $\delta^{13}C$-values were measured over most of the incubation time, but after 42 days, the signal was strongly reduced. Additionally to the isotopic

Table 8.1 Isotopic composition ($\delta^{13}C$) and total organic carbon content (TOC) of the cytoplasm of living foraminifera (*Ammonia tepida*=*A. tepida*, *Bolivina variabilis*=*B. variabilis*) incubated for different time intervals

Species	Days	$\delta^{13}C$ (‰)	TOC (mgC)
A. tepida	0	−16.39	0.2675
A. tepida	2	2,839.94	0.2361
A. tepida	7	1,934.50	0.2107
A. tepida	21	1,464.89	0.1908
A. tepida	42	1,101.30	0.1655
B. variabilis	0	−17.33	0.0629
B. variabilis	2	11,147.02	0.0759
B. variabilis	7	14,732.92	0.0446
B. variabilis	21	9,241.48	0.0776
B. variabilis	42	8,829.89	0.0665

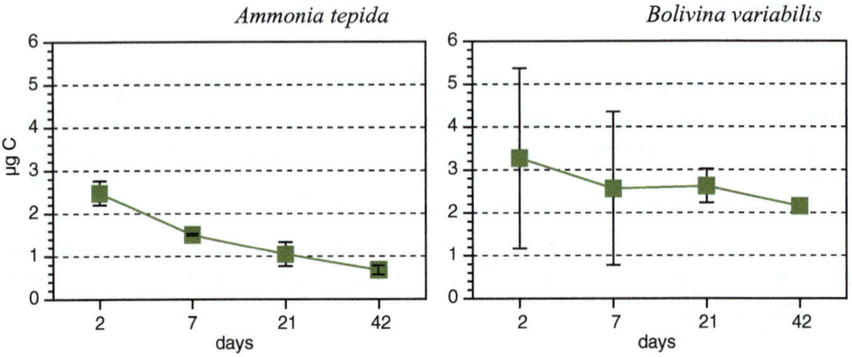

Fig. 8.6 Cytoplasmic carbon uptake (µg C) by *A. tepida* and *B. variabilis* incubated for different time intervals

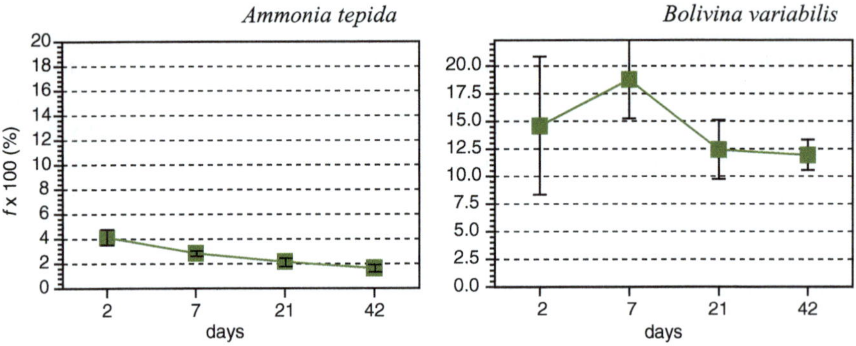

Fig. 8.7 Fraction f of carbon originated from ^{13}C-labeled *Dunaliella tertiolecta* and found in the cytoplasm of *A. tepida* and *B. variabilis* incubated for different time intervals

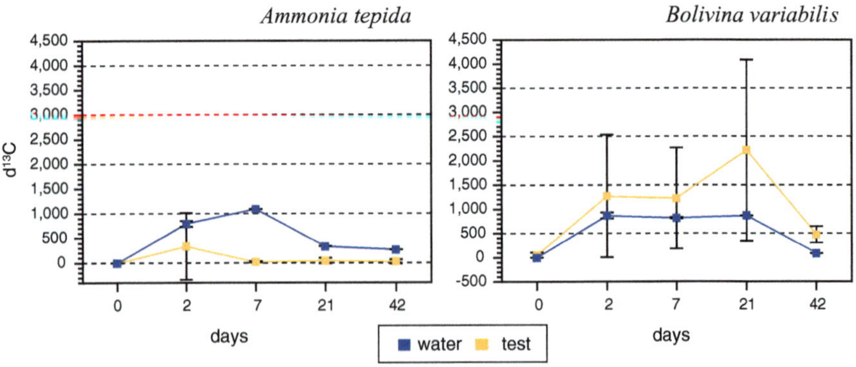

Fig. 8.8 Isotopic composition ($\delta^{13}C$) of water and foraminiferal carbonate tests of *A. tepida* and *B. variabilis* incubated for different time intervals

signal in the overlaying water, the isotopic signal in the foraminiferal tests is shown in Fig. 8.8. For *A. tepida*, slightly elevated δ^{13}C-values in the tests were recorded just after 2 days. Afterwards, tests shifted back to lighter values. For *B. variabilis*, δ^{13}C-values in the test increased with incubation time up to 21 days and got strongly reduced afterwards. But again, high variability was seen between the replicates and a high standard deviation is visible for the isotopic signal in the tests of *B. variabilis.*

8.4 Discussion

8.4.1 Food Source

^{13}C-labeled *D. tertiolecta* was used as isotopic marker for food ingestion in our experiments. It is a unicellular algae belonging to the *Chlorophyceae. Dunaliella* has been used a representative of phytoplankton by previous studies and culture experiments (e.g., Marechal-Abram et al. 2004; Bernhard et al. 2009; Filipsson et al. 2010) and the fast and active response of foraminifera to *D. tertiolecta* has been reported from the laboratory (Heinz et al. 2002) as well as from field studies in Sagami Bay (Kitazato et al. 2003; Nomaki et al. 2005a, 2006).

8.4.2 Foraminiferal Uptake

After an single and initial food pulse at the beginning (t=0) of two experimental time series, the uptake of labeled food was maximum in samples incubated for 2 days. This trend was the same for *A. tepida* as well *as B. variabilis* (Fig. 8.6). Longer incubation times reduced the uptake signal in both series. This indicates that the ingestion of food material and *D. tertiolecta* by the foraminifera was maximum during the initial 2 days of incubation and it steadily decreased thereafter. The fast uptake of labeled food by the foraminifers shows their affinity for fresh organic matter. With ongoing time, the amount of available food surely decreased because of foraminiferal consumption. Additionally, it is expected that with time, the labeled algal material started to degrade. Previous ^{13}C labeling experiments using algal material found that certain parts of the added algae were degraded after a few hours to days (Middelburg et al. 2000; Thomas and Blair 2002; Moodley et al. 2005). Both processes, fast foraminiferal consumption at the beginning of the experiment and constant algae degradation, probably led to reduced availability of labeled food later in the time series and weakening uptake rates with time.

In the *A. tepida* series, 350 individuals ingested in total 2.5 μgC after 2 days of incubation, indicating that, on average, each individual ingested 7.14 ngC. After 42 days, each *A. tepida* individual still ingested 2 ngC. For *B. variabilis* carbon uptake ranges between 9.43 ngC/individual after 2 days and 6.23 ngC/individual after 42 days. Experimental dishes were 7.4-cm diameter and, therefore, provided an area of 43 cm^2. For comparison of the direct carbon uptake of both series with in situ investigations, recalculations are necessary: *A. tepida* recorded values between 0.58 mgC/m^2 after 2 days and 0.16 mgC/m^2 after 42 days, *B. variabilis* ranged between 0.77 and 0.51 mgC/m^2. Our experimental carbon uptake rates seems therefore to be very low, within a level comparable to abyssal in situ experiments including foraminifera and other meiofauna (Sweetman and Witte 2008b; Ingels et al. 2011; Enge et al. 2011) or ex situ experiments directly on board (Moodley et al. 2005; Sweetman and Witte 2008a; Sweetman et al. 2009). Bathyal to intertidal in situ experiments and some ex situ intertidal experiments show typically levels higher than 1.5 mgC/m^2, up to 31 mgC/m^2 (Nomaki et al. 2005a; Witte et al. 2003a, b; Moodley et al. 2002, 2005; Hunter et al. 2012). But this direct comparison is unrealistic because field studies analyze the total fauna in a specific area, and usually total foraminiferal abundances are significantly higher compared to our isolated 350 individuals.

8.4.3 Species-Specific Carbon Ingestion

Bolivina variabilis recorded higher uptake values than *A. tepida*, indicating that the extent of foraminiferal response to labeled food is species specific. Both species considerably differ in size and biomass, which has of course an influence on the species-specific amount of uptake. Therefore, the comparison of the fraction of carbon originating from added algae (Fig. 8.7) can sometimes be more useful that a direct comparison of uptake rates. Here, the higher fraction f is clearly visible in *B. variabilis*. After 2 days, about 15 % of the carbon in the cytoplasm of *B. variabilis* derived from the added food material, after 7 days it is even more than 18 %. Comparable high amounts after 2–6 days incubation with *D. tertiolecta* as food source were reported during in situ experiments *for Uvigerina akitaensis*, *Bulimina aculeata* and *Bolivina pacifica* from Sagami Bay (Nomaki et al. 2005a). All of them actively ingest fresh algal material and quickly react to algae deposition. Members of the genus *Uvigerina* and *Bulimina* are important indicators for high productivity (Loubere and Fariduddin 1999; Fontanier et al. 2003) and *B. variabilis* analyzed in our investigations shows similar characteristics. In our cultures, individuals of *B. variabilis* are strongly associated to algae mats (Fig. 8.3a, b) and prefer this kind of habitat on the sediment surface. Based on the results, we attribute this behavior of *B. variabilis* to their preference to high amounts of available fresh food.

Although the signal in *A. tepida* is weaker, about 4 % of the carbon derived from the added food material after 2 days, which is comparable to the signal in *Textularia kattegatensis* during in situ experiments in Sagami Bay (Nomaki et al. 2005a). *Ammonia tepida* is reported to be a herbivore, not a detritivore species (Murray 1991; Burone et al. 2007); our results support this assertion.

8.4.4 *Isotopic Signal in the Water*

In both time series, an increase in $\delta^{13}C$-values in the water caused by decomposition processes of the ^{13}C-labeled algae were detected at the beginning of incubation (Fig. 8.8). Decomposition of food can result from bacterial activity. Additionally, foraminifera degrade the food material in the dish by ingestion, digestion and respiration. A considerable amount of the algal material may be excreted after digestion and the rate of excretion varies in different species. Foraminifera can excrete more than 50 % of the carbon derived from the food within 24 h as previously observed in the case of *A. lobifera* (Kuile et al. 1987). For *A. tepida*, the water signal and, therefore, decomposition processes seem to decrease after a maximum of 7 days. For *B. variabilis*, values do not decrease until the end of the incubation, after 42 days.

8.4.5 *Isotopic Signal in Foraminiferal Tests*

Figure 8.8 recorded the isotopic signals measured during the analysis of the foraminiferal tests. The isotopic signal in the test should mirror the isotopic signal of the water mass during calcification of this test (e.g. Saraswat 2006). It would be interesting to estimate the amount of algal carbon that is finally manifested in the foraminiferal test after ingestion of food material. Kuile et al. (1987) observed that less than 5 % of the carbon taken up by the foraminifera through feeding is incorporated in to the tests; the rest is retained in the organic fraction. Our results show elevated $\delta^{13}C$-values in the tests at the beginning of both time series. But unfortunately, in both series values decline afterwards, for *A. tepida* quickly after 7 days, for *B. variabilis* just at the end of incubation after 42 days. It is not clear what caused this reduced signal in the tests in both series. One possible explanation is that high calcification took place during the incubation. At the beginning, elevated $\delta^{13}C$-values in the water, caused by decomposition processes, could lead to elevated $\delta^{13}C$-values in the tests. Later on, heavy ^{13}C isotopes in the water get reduced and new calcification processes could diminish the signal in the tests. But for *A. tepida*, $\delta^{13}C$-values in the tests were already reduced at 7 days, although water samples still showed high ratios (Fig. 8.8). Additionally, no significant growth of the tests was observed during incubation time, and most of the added foraminifera were judged as adults from the very beginning because of their size. Decalcification processes during incubation would also explain the signal lost in the tests at the end of the series, but tests looked intact and no signs of decalcification were visible with our binocular stereomicroscope. SEM examinations would be required to judge about this.

The most likely explanation seems to be problems during sample preparation and measurement. Despite of cleaning processes, it was not possible to remove the cytoplasm in the tests totally, and remnants of organic material were recognizable, which may lead to contamination during test measurements. For measuring the isotopic signal of foraminiferal calcareous tests, shells were dissolved with phosphoric acid (H_3PO_4) and the released carbon dioxide will be measured afterwards in the mass spectrometer. But the phosphoric acid normally should not affect organic material, and will not release CO_2 from there. But we cannot exclude that inorganic particles were stored in the cytoplasm, released by the acid and finally contaminated the analysis.

Acknowledgements We thank Emmanuelle Geslin and her coworkers from the Laboratoire d`Etude des Bio-indicateurs Actuels et Fossiles (BIAF), University of Angers, France as well as our colleague Annekatrin Enge for the great support during sample collection in the field. Thanks a lot to our collegue Roman Marten for all his help when picking the foraminifera. We are very grateful to Ulrich Struck from the Museum für Naturkunde in Berlin for analyzing all our isotopic samples and his helpful advice concerning cleaning foraminiferal tests. Authors LVN and RN thank the director of NIO, Goa for the necessary support; LVN specially thank UGC-DAAD for facilitating the visit to Germany and the Department of Science and Technology for the financial support in the form of Fast Track young scientist fellowship.

References

Altenbach AV (1992) Short term processes and patterns in the foraminiferal response to organic flux rates. Mar Micropal 19:119–129

Altenbach AV, Struck U (2001) On the coherence of organic carbon flux and benthic foraminiferal biomass. J Foramin Res 31:79–85

Altenbach AV, Pflaumann U, Schiebel R, Thies A, Timm S, Trauth M (1999) Scaling percentages and distributional patterns of benthic foraminifera with flux rates of organic carbon. J Foramin Res 29:173–185

Andersson JH, Woulds C, Schwartz M, Cowie GL, Levin LA, Soetaert K, Middelburg JJ (2008) Short-term fate of phytodetritus in sediments across the Arabian Sea oxygen minimum zone. Biogeosciences 5:43–53

Berger WH, Smetacek V, Wefer G (1989) Ocean productivity and paleoproductivity: an overview. In: Berger WH, Smetacek V, Wefer G (eds) Productivity of the ocean: present and past, Dahlem workshop reports. Wiley, Chichester, pp 1–34

Bernhard JM, Mollo-Christensen E, Eisenkolb N, Starczak VR (2009) Tolerance of allogromid foraminifera to severely elevated carbon dioxide concentrations: implications to future ecosystem functioning and paleoceanorgaphic interpretations. Global Planet Change 65: 107–114

Blair NE, Levin LA, DeMaster DJ, Plaia G (1996) The short-term fate of fresh algal carbon in continental slope sediments. Limnol Oceanogr 41:1208–1219

Burone L, Valente P, Pires-Vanin A, Mello e Souza SH, Mahiques MM, Braca E (2007) Benthic foraminiferal variability on a monthly scale in a subtropical bay moderately affected by urban sewage. Scient Mar 71:775–792. doi:10.3989/scimar.2007.71n4775

Corliss BH, Emerson S (1990) Distribution of Rose Bengal stained deep-sea benthic foraminifera from the Nova Scotian continental margin and Gulf of Maine. Deep-Sea Res 37:381–400

Enge A, Nomaki H, Ogawa N, Witte U, Moeseneder M, Lavik G, Ohkouchi N, Kitazato H, Kucera M, Heinz P (2011) Response of the benthic foraminiferal community to a simulated short-term phytodetritus pulse in the abyssal North Pacific. Mar Ecol Prog Ser 438:129–142

Filipsson HL, Bernhard JM, Lincoln SA, McCorkle DC (2010) A culture-based calibration of benthic foraminiferal paleotemperature proxies: [delta]18O and Mg/Ca results. Biogeosciences 7:1335–1347

Fontanier C, Jorissen FJ, Licari L, Alexandre A, Anschutz P, Carbonel P (2002) Live benthic foraminiferal faunas from the Bay of Biscay: faunal density, composition, and microhabitats. Deep-Sea Res I 49:751–785

Fontanier C, Jorissen FJ, Chaillou G, David C, Anschutz P, Lafon V (2003) Seasonal and interannual variability of benthic foraminiferal faunas at 550 m depth in the Bay of Biscay. Deep-Sea Res I 50:457–494

Giullard RR, Ryther JH (1962) Studies of marine planktonic diatoms. I. *Cyclotella nana* Hustedt and *Detonula confervacea* (Cleve). Gran Can J Microbiol 8:229–239

Gooday AJ (1993) Deep-sea benthic foraminiferal species which exploit phytodetritus: characteristic features and controls on distribution. Mar Micropaleontol 22:187–205

Gooday AJ, Lambshead PJD (1989) Influence of seasonally deposited phytodetritus on benthic foraminiferal populations in the bathyal northeast Atlantic: the species response. Mar Ecol Prog Ser 58:53–67

Gooday AJ, Rathburn AE (1999) Temporal variability in living deep-sea benthic foraminifera: a review. Earth Sci Rev 46:187–212

Gooday AJ, Turley CM (1990) Response by benthic organisms to inputs of organic material to the ocean floor: a review. Philos Trans R Soc Lond A 331:119–138

Gooday AJ, Levin LA, Linke P, Hegger T (1992) The role of benthic foraminifera in deep-sea food webs and carbon cycling. In: Rowe GT, Pariente V (eds) Deep-sea food chains and global carbon cycle. Kluwer, The Netherlands, pp 63–91

Heinz P, Kitazato H, Schmiedl G, Hemleben C (2001) Response of deep-sea benthic foraminifera from the Mediterranean Sea to simulated phytoplankton pulses under laboratory conditions. J Foramin Res 31:210–227

Heinz P, Hemleben C, Kitazato H (2002) Time-response of cultured deep-sea benthic foraminifera to different algal diets. Deep-Sea Res I 49(517–537):107

Heip CHR, Duineveld G, Flach E, Graf G, Helder W, Herman PMJ, Lavaleye M, Middelburg JJ, Pfannkuche O, Soetaert K, Soltwedel T, de Stigter H, Thomsen L, Vanaverbeke J, de Wilde P (2001) The role of the benthic biota in sedimentary metabolism and sediment-water exchange processes in the Goban Spur area (NE Atlantic). Deep-Sea Res II 48:3223–3243

Herguera JC (1992) Deep-sea benthic foraminifera and biogenic opal: glacial to post glacial productivity changes in the western equatorial Pacific. Mar Micropaleontol 19:79–98

Hunter W, Levin L, Kitazato H, Witte U (2012) Macrobenthic assemblage structure and organismal stoichiometry control faunal processing of particulate organic carbon and nitrogen in oxygen minimum zone sediments. Biogeosciences 9:993–1006

Ingels J, Billett DSM, Van Gaever S, Vanreusel A (2011) An insight into the feeding eecology of deep-sea canyon nematodes: results from field observations and the first *in-situ* ^{13}C feeding experiment in the Nazaré Canyon. J Exp Mar Biol Ecol 396:185–193

Jorissen FJ, Barmawidjaja BM, Puskaric S, van der Zwaan GJ (1992) Vertical distribution of benthic foraminifera in the northern Adriatic Sea: the relation with the organic flux. Mar Micropaleontol 19:131–146

Jumars PA, Mayer LM, Deming JW, Baross JA, Wheatcroft RA (1990) Deep-sea deposit- feeding strategies suggested by environmental and feeding constraints. Phil Trans Roy Soc London A 331:85–101

Kitazato H, Shirayama Y, Nakatsuka T, Fujiwara S, Shimanaga M, Kato Y, Okada Y, Kanda J, Yamaoka A, Masuzawa T, Suzuki K (2000) Seasonal phytodetritus deposition and responses of bathyal benthic foraminiferal populations in Sagami Bay, Japan: preliminary results from 'Project Sagami 1996–1999'. Mar Micropaleontol 40:135–149

Kitazato H, Nomaki H, Heinz P, Nakatsuka T (2003) The role of benthic foraminifera in deep-sea food webs at the sediment–water interface: results from *in situ* feeding experiments in Sagami Bay. Front Res Earth Evol 1:227–232

Kuile B ter, Erez J, Lee JJ (1987) The role of feeding in the metabolism of larger symbiont bearing foraminifera. Symbiosis 4:335–336

Levin LA, Blair NE, Martin CM, DeMaster DJ, Plaia G, Thomas CJ (1999) Macrofaunal processing of phytodetritus at two sites on the Carolina margin: *in situ* experiments using ^{13}C-labeled diatoms. Mar Ecol Prog Ser 182:37–54

Loubere P, Fariduddin M (1999) Benthic foraminifera and the flux of organic carbon to the seabed. In: Sen Gupta BK (ed) Modern foraminifera. Kluwer, Dordrecht, pp 181–199
Marechal-Abram N, Debenay J-P, Kitazato H, Wadai H (2004) Cadmium partition coefficients of cultured benthic foraminifera *Ammonia beccarii*. Geochem J 38:271–283
Miao Q, Thunell RC (1993) Recent deep-sea benthic foraminiferal distributions in the South China and Sulu Seas. Mar Micropaleontol 22:1–32
Middelburg JJ, Barranguet C, Boschker HTS, Herman PMJ, Moens T, Heip CHR (2000) The fate of intertidal microphytobenthos carbon: an *in situ* ^{13}C-labeling study. Limnol Oceanogr 45:1224–1234
Moodley L, Boschker HTS, Middelburg JJ, Pel R, Herman PMJ, de Deckere E, Heip CHR (2000) Ecological significance of benthic foraminifera: ^{13}C labelling experiments. Mar Ecol Prog Ser 202:289–295
Moodley L, Middelburg JJ, Boschker HTS, Duineveld GCA, Pel R, Herman PMJ, Heip CHR (2002) Bacteria and foraminifera: key players in a short-term deep-sea benthic response to phytodetritus. Mar Ecol Prog Ser 236:23–29
Moodley L, Middelburg JJ, Soetaert K, Boschker HTS, Herman PMJ, Heip CHR (2005) Similar rapid response to phytodetritus deposition in shallow and deep-sea sediments. J Mar Res 63:457–469
Murray JW (1991) Ecology and palaeoecology of benthic foraminifera. London, Longman Scientific & Technical, Harlow, U.K, 397 p
Nomaki H, Heinz P, Nakatsuka T, Shimanaga M, Kitazato H (2005a) Species-specific ingestion of organic carbon by deep-sea benthic foraminifera and meiobenthos: In situ tracer experiments. Limnol Oceanogr 50:134–146
Nomaki H, Heinz P, Nakatsuka T, Shimanaga M, Ohkouchi N, Ogawa NO, Kogure K, Ikemoto E, Kitazato H (2006) Different ingestion patterns of ^{13}C-labeled bacteria and algae by deep-sea benthic foraminifera. Mar Ecol Prog Ser 310:95–108
Nomaki H, Ohkouchi N, Heinz P, Suga H, Chikaraishi Y, Ogawa NO, Matsumoto K, Kitazato H (2009) Degradation of algal lipids by deep-sea benthic foraminifera: an *in situ* tracer experiment. Deep-Sea Res Part I 56:1488–1503
Ohga T, Kitazato H (1997) Seasonal changes in bathyal foraminiferal populations in response to the flux of organic matter (Sagami Bay, Japan). Terra Nova 9:33–37
Pedersen TF, Pickering M, Vogel JS, Southon JN, Nelson DE (1988) The response of benthic foraminifera to productivity cycles in the eastern equatorial Pacific: faunal and geochemical constraints on glacial bottom water oxygen levels. Paleoceanography 3:157–168
Pfannkuche O (1993) Benthic response to the sedimentation of particulate organic matter at the BIOTRANS station, 47°N, 20°W. Deep-Sea Res II 40:135–149
Rathburn AE, Corliss BH (1994) The ecology of living (stained) benthic foraminifera from the Sulu Sea. Paleoceanography 9:87–150
Saraswat R (2006) Development and evaluation of proxies for high-resolution paleoclimate reconstruction with special reference to northern Indian ocean. Dissertation, Goa University
Shimanaga M, Kitazato H, Shirayama Y (2000) Seasonal patterns of vertical distribution between meiofaunal groups in relation to phytodetritus deposition in the bathyl Sagami bay, central Japan. J Oceanogr 56:379–387
Smith KL, Kaufmann RS, Baldwin RJ (1994) Coupling of near-bottom pelagic and benthic processes at abyssal depths in the eastern north Pacific Ocean. Limnol Oceanogr 39:1101–1118
Snider LJ, Burnett BR, Hessler RR (1984) The composition and distribution of meiofauna and nanobiota in a central North Pacific deep-sea area. Deep-Sea Res 31:1225–1249
Sweetman AK, Witte U (2008a) Macrofaunal response to phytodetritus in a bathyal Norwegian fjord. Deep-Sea Res I 55:1503–1514
Sweetman AK, Witte U (2008b) Response of an abyssal macrofaunal community to a phytodetrital pulse. Mar Ecol Prog Ser 355:73–84
Sweetman AK, Sommer S, Pfannkuche O, Witte U (2009) Retarded response by macrofauna-size foraminifera to phytodetritus in a deep Norwegian fjord. J Foramin Res 39:15–22

Thomas CJ, Blair NE (2002) Transport and digestive alteration of uniformly ^{13}C-labeled diatoms in mudflat sediments. J Mar Res 60:517–535

Wahyudi C, Minagawa M (1997) Response of benthic foraminifera to organic carbon accumulation rates in the Okinawa trough. J Oceanogr 53:411–420

Witte U, Aberle N, Sand M, Wenzhofer F (2003a) Rapid response of a deep-sea benthic community to POM enrichment: an *in situ* experimental study. Mar Ecol Prog Ser 251:27–36

Witte U, Wenzhofer F, Sommer S, Boetius A, Heinz P, Aberle N, Sand M, Cremer A, Abraham WR, Jorgensen BB, Pfannkuche O (2003b) *In situ* experimental evidence of the fate of a phytodetritus pulse at the abyssal sea floor. Nature 424:763–766

Woulds C, Cowic GL, Lcvin LA, Andcrsson JII, Middelburg JJ, Vandewiele S, Lamont PA, Larkin KE, Gooday AJ, Schumacher S, Whitcraft C, Jeffreys RM, Schwartz M (2007) Oxygen as a control on seafloor biological communities and their roles in sedimentary carbon cycling. Limnol Oceanogr 52:1698–1709

Chapter 9
How Has Foraminiferal Genetic Diversity Developed? A Case Study of *Planoglabratella opercularis* and the Species Concept Inferred from Its Ecology, Distribution, Genetics, and Breeding Behavior

Masashi Tsuchiya, Kenji Takahara†, Mutsumu Aizawa, Hitomi Suzuki-Kanesaki, Takashi Toyofuku, and Hiroshi Kitazato

Abstract The nature of *Planoglabratella opercularis* has been revealed through multidisciplinary studies focusing on its ecology, life cycle, growth rate, interbreeding, morphology, natural abundance, distribution, and molecular ecology. Our research is aimed at revealing the processes and mechanisms behind foraminiferal evolution. *P. opercularis* is a benthic inhabitant of rocky-shore environments, where it crawls on coralline algae. Gametes are not released into the ambient seawater; rather this species forms plastogamic pairs and exchanges gametes inside the shell. Thus, the mobility and dispersal of both individuals and gametes are extremely low. In fact, genetic divergence has likely occurred within each geographic population. Interpopulational breeding experiments revealed breeding incompatibility between gamont specimens from widely separated local populations. The genetic connectivity in *P. opercularis* is affected by ecological characteristics such as habitat and mobility. Molecular ecological studies have revealed that *P. opercularis* likely diversified through reproductive isolation resulting from geographical separation following changes in land–ocean distributions.

Keywords Foraminifer culture • Breeding population • Intraspecific genetic variation • Molecular phylogeny • Speciation • Species concept

M. Tsuchiya (✉) • T. Toyofuku • H. Kitazato
Japan Agency for Marine-Earth Science and Technology (JAMSTEC), Institute of Biogeosciences, 2-15 Natsushima-cho, Yokosuka 237-0061, Japan
e-mail: tsuchiyam@jamstec.go.jp

K. Takahara†
Deceased June 2, 2013

M. Aizawa • H. Suzuki-Kanesaki
Faculty of Science, Shizuoka University, 836 Oya, Shizuoka 422-8529, Japan

H. Kitazato and J.M. Bernhard (eds.), *Approaches to Study Living Foraminifera: Collection, Maintenance and Experimentation*, Environmental Science and Engineering, DOI 10.1007/978-4-431-54388-6_9, © Springer Japan 2014

9.1 Introduction

In this chapter we address the questions of how foraminiferal genotypic diversification occurred, and how foraminiferal biodiversity developed. The ecological and biological meanings of genetic differences within and among phylogenetic lineages are not yet clearly understood, in part because of discrepancies between morphologic and molecular data. When we use fossil records, we have to distinguish "species" based only on morphology. However, morphology is flexible and can be changed by the ambient environmental conditions. For example, in foraminifera the development of spines and the arrangement of chambers are controlled genetically, but pore size and density, length of spines, and shell height are influenced by the ambient environment (Tsuchiya et al. 2008, 2009). It is necessary to separate ecophenotypic variants and genetically stable characters to prevent incompatibility between morphologic and molecular data.

What defines a species of Foraminifera? In other words, how are genotypes distinguished? Before populational genetics can be applied, the meaning of "species" should be made clear. To date there have been many attempts to recognize genotypes in both planktonic and benthic foraminifers (e.g., Darling and Wade 2008; Holzmann 2000). One result of these studies is the recognition that genetic exchange may occur between Arctic and Antarctic oceans within certain planktonic and benthic foraminifers (Darling et al. 2004; Pawlowski et al. 2007). However, there is as yet no consensus as to the difference between a genotype and a biological species as a breeding population.

Planoglabratella opercularis (d'Orbigny) is a common foraminiferal species (Fig. 9.1) widely distributed in the intertidal zone of rocky shores around the Japanese Islands that is currently the focus of multidisciplinary research. This species is easy

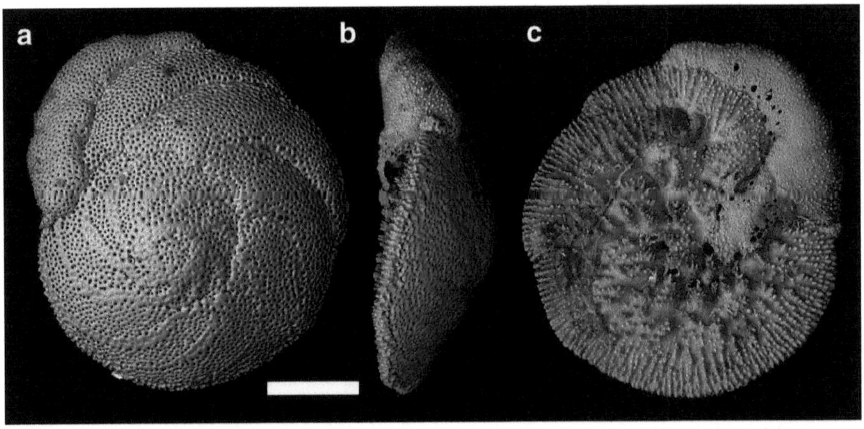

Fig. 9.1 Scanning electron micrograph of specimen of *Planoglabratella opercularis*. Dorsal view (**a**), side view (**b**) and ventral view (**c**) of the same specimen. Scale bar, 100 μm

to culture in the laboratory and its morphological variations, reproductive behavior, and genetics are well documented (Kitazato 1992; Kitazato et al. 2000; Tsuchiya et al. 2000, 2003). We have been studying the ecology of this species, making ultrastructural observations, delineating breeding populations, investigating its molecular ecology and phylogeography using nuclear-encoded ribosomal RNA (rRNA) genes, and performing global phylogenetic studies of rhizarian lineages using protein sequences with this species. Specific research involving *P. opercularis* includes establishing its distribution and life cycle, inferred from both culture experiments and field observations (Aizawa 1987, unpublished data; Kanesaki 1987, unpublished data; Tsuchiya et al. 1994); re-evaluation of its morphology and estimation of interpopulational structure, based on breeding experiments (Takahara 1989, unpublished data; Kitazato et al. 2000); molecular phylogenetic analysis of glabratellids and local populations of *P. opercularis*, based on small subunit (SSU) and large subunit (LSU) rRNA genes, and the internal transcribed spacer region (ITS) of nuclear rRNA gene sequences (Tsuchiya et al. 2000, 2003); evaluation of Mg/Ca thermometry and isotope analyses, based on culture and field observations (Toyofuku et al. 2000); determination of the relationship between foraminifera and cercozoans, based on tubulin and actin phylogeny (Takishita et al. 2005); and examination of the global phylogeny within the rhizarian lineage, by using several amino acid sequences (Kamikawa et al. 2008; Sakaguchi et al. 2009; Ishitani et al. 2011, 2012).

To assess speciation in the Foraminiferida, it is important to combine information from studies of foraminiferal ecology, morphology, molecular phylogeny, genetics, and geographic distribution (Murray 2006), not only to determine true taxonomic relationships but also to distinguish both diversification and evolutionary processes. Moreover, the meaning of a clade (lineage) must be clarified, and the means to distinguish genotypes must also be established. A multidisciplinary approach using *P. opercularis* as a model species can be used to address these issues.

9.2 Ecology, Distribution and Natural Abundance of *P. opercularis*

Planoglabratella opercularis (Glabratellidae), a benthic foraminifer, is distributed on rocky shores along the Japanese coast, crawling on the thalli of coralline algae and grazing on microalgae or organic detritus on the surface of seaweeds (Kitazato 1984, 1988, 1994). As with *Glabratella sulcata*, *P. opercularis* is thought to have a biphasic life cycle with sexual and asexual stages that show a gamontogamic–paraclassical life cycle (Grell 1958, 1979; Lee et al. 1991). During sexual reproduction, two or more individuals form a plastogamic pair and exchange gametes within the shell. Because of their behavior, the dispersal of both the gametes and individuals is thought to be extremely low.

There are five major species of Glabratellidae distributed in different patterns around the Japanese Islands: *Glabratella patelliformis*, *G. milletti*, *P. nakamurai*,

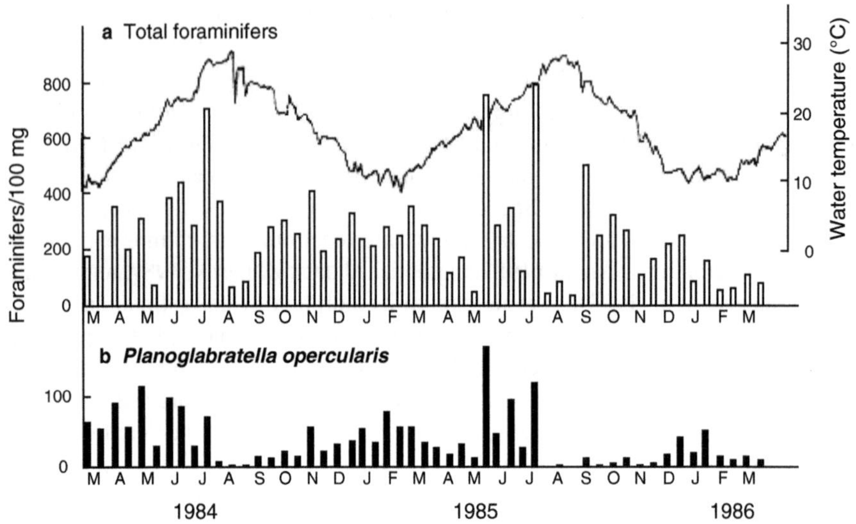

Fig. 9.2 Seasonal fluctuations in numbers of foraminifera in coralline algae samples. Total foraminifers (**a**) and *P. opercularis* (**b**) at Omaezaki, Shizuoka Prefecture, Japan

P. opercularis, and *Angulodiscorbis quadrangularis*. The distribution patterns are reportedly influenced by water temperatures (Kanesaki 1987, unpublished data). *P. opercularis* is distributed mainly in the northern part of the Japanese Islands, showing temperate to cold water-associated distribution. In contrast, *P. nakamurai* has a patchy distribution over a limited area around the Japanese Islands; around Shikoku Island and along the coast from Izu Peninsula to Boso Peninsula. Field observations suggest that water temperatures affect the distribution of these species. It is also possible that the distribution patterns of these species are related to their evolutionary history.

Both the life history and the life span of *P. opercularis* were studied through serial observations at one-month intervals from 1984 to 1986 at a fixed station at Omaezaki (35°35.7 N, 138°13.6 E), Shizuoka, Japan (Fig. 9.2) (Aizawa 1987, unpublished data). The frequency distribution of the shell diameters shows a broad range from small juveniles to adults through the seasons; juvenile specimens with around ten chambers were present year-round, especially in spring and autumn (Fig. 9.3). In *P. opercularis*, the alternation of generations is distinguishable through morphological differences of coiling direction (the asexual generation has the sinistral form and the sexual generation has the dextral form; detailed in Sect. 9.4.3) and maximum diameter of the specimens (the sexual dextral form, up to approximately 250 μm; the asexual sinistral form, up to approximately 500 μm). Both morphologically distinguishable generations are found in any season, suggesting that *P. opercularis* has a short life span, co-existed different growth stage from small juveniles to adults. It is difficult to infer the population dynamics of natural populations from the available size-distribution data.

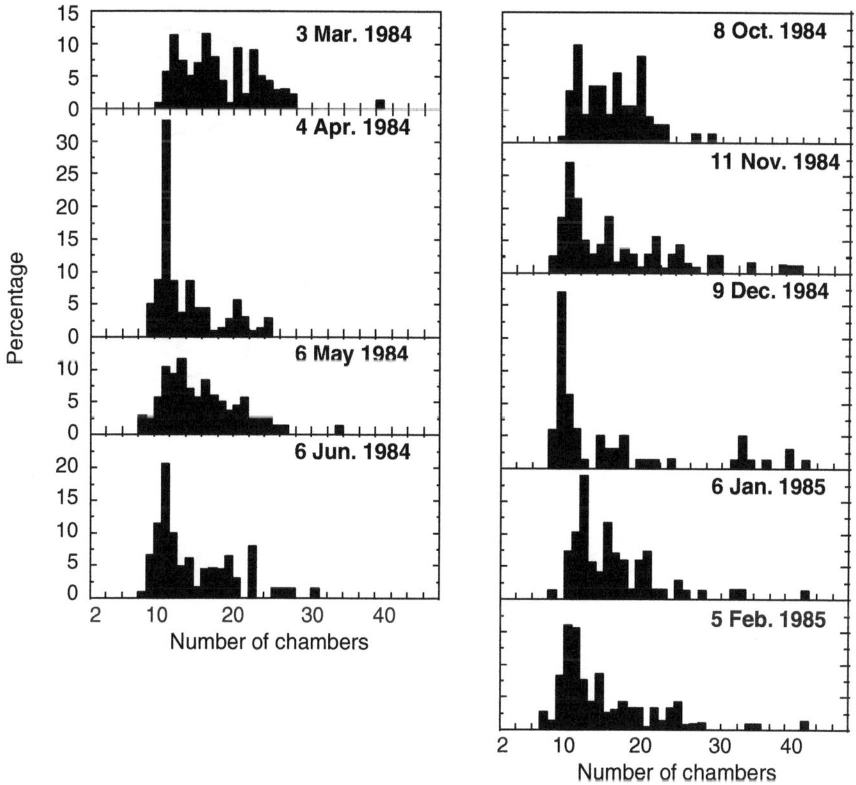

Fig. 9.3 Size distributions of *P. opercularis* collected at Omaezaki, Japan. Juvenile specimens were commonly found in this area at any seasons

9.3 Culture Experiments with *P. opercularis*

9.3.1 Culturing Methods for Life Cycle Characterization, Clone Cultures, and Breeding Experiments

To characterize the life cycle of *P. opercularis*, we conducted culture experiments and estimated growth rates. Samples for culture experiments were collected from Omaezaki. Living *P. opercularis* specimens attached to coralline algae were collected from tide pools and placed in large culture tanks filled with well-oxygenated seawater as a "rough culture" (Kitazato 1984, 1988). Living foraminifers were transferred from these culture tanks into a small Petri dish (raw culture). A total of 619 individuals were used for growth-rate observations and 1,070 individuals (535 pairs) were used for breeding experiments. Water for both raw and clone cultures was collected from the collecting site and sterilized by filtration through 0.2-μm

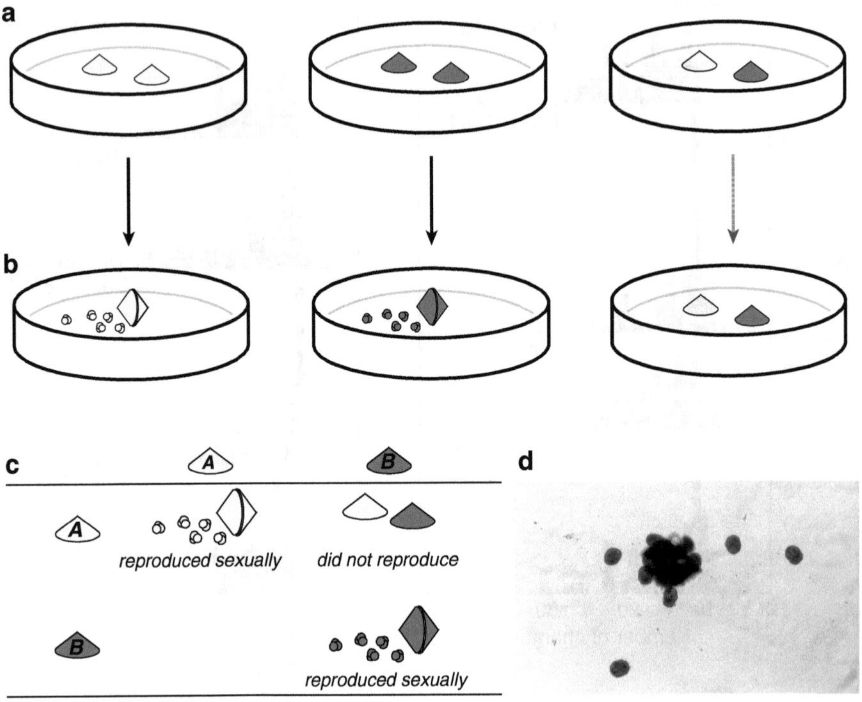

Fig. 9.4 Concept behind breeding experiments. Two specimens from the same or different locations (*white* and *gray*) are placed in a Petri dish (**a**) until a plastogamic pair forms (**b**). In the example shown, individuals from same location can breed and sexually reproduce, but individuals from different locations did not interbreed (**c**). Sexual reproduction of *P. opercularis* (**d**)

pore-size membrane filters. To maintain the dissolved oxygen levels (approximately 6.0 mL/L), culture water was changed by pipette every day. Specimens were kept in Petri dishes in a temperature-controlled incubator. Light conditions in the incubator were a 12:12-h light:dark cycle and a luminance of about 4,000 lx. Temperatures were set at 10, 15, 20, 25 and 30 °C, and salinities were 25, 30, 35, and 40 psu.

Growth rates were determined by observing three-chambered juvenile offspring, produced sexually or asexually. Individual three-chambered juvenile specimens are very fragile, so they were carefully handled with a thin flexible needle.

The breeding capability of glabratellid species can be directly observed. When two individual specimens placed in the same Petri dish formed a plastogamic pair and reproduced sexually, they were considered to be from the same breeding population (Fig. 9.4). This classic method is simple, but it is one of the indices for determining a breeding population.

We used the same culture system as that used for growth-rate estimates by Kitazato et al. (2000), who also conducted interspecific breeding experiments with three morphospecies, *P. nakamurai*, *P. opercularis*, and *G. patelliformis*. Two individuals collected from different localities were placed in a Petri dish with seawater and observed

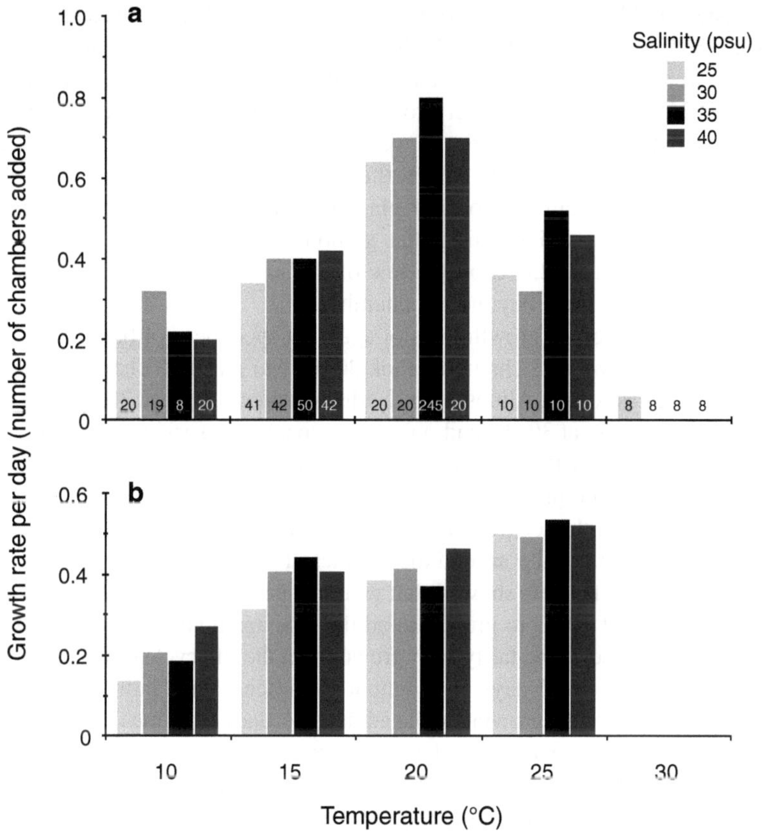

Fig. 9.5 Growth rate of juvenile gamont specimens at different water temperatures and salinities. We used three-chambered juveniles at the start of culture experiments. Average growth rates are shown as the number of chambers added per day for 5 days (**a**) or for 14 days (**b**) since the experiment began. Numbers in each column indicate the initial number of individual specimens observed (**a**)

until they formed a plastogamic pair. Plastogamic pair formation was observed by using a phase-contrast apparatus attached to an inverted microscope (Nikon-TMD Cultivation Microscope System and Olympus IMT-2 Inverted Microscope System, Tokyo, Japan). The gamontogamy process was recorded by an automatic microphotographic system (Nikon-HFM) attached to the inverted microscope.

9.3.2 Growth Rate and Life Cycle

Average growth rates were estimated from both clone cultures of gamont specimens and sexually produced offspring of agamont specimens of *P. opercularis* under controlled environmental conditions (Fig. 9.5). Juvenile gamonts added an average of 3.2–4.0 chambers over the first 5 days (0.64–0.80 chambers/day), and grew fastest at 20 °C and 35 psu, of the temperatures and salinities tested. Growth rates at 25 °C were lower than at 20 °C, and there was no growth and only a few survivors at any salinity at 30 °C. These results suggest that continuous temperatures over 25 °C negatively affect growth and survival rates of the gamont stage. Slight differences in the growth rate were evident after 14 days at 15 and 20 °C, and growth was slow at 10 °C. When they reached a size of about 15 chambers (about 250 μm), specimens formed plastogamous pairs. Very few individuals grew beyond this size, indicating that gamonts reach the adult stage in about 15 days. In contrast, at 25 °C only a few of the surviving specimens grew to ten chambers and none grew beyond ten chambers.

We measured average growth rates of agamont specimens of both juveniles and adults. Specimens with 15 chambers took 30 days to reach 30 chambers, and then they reproduced asexually. Almost all individuals cannot survive at 30 °C for a culture period, died instantly at 30 °C, and only a few specimens survived at 25 °C.

The distribution of this species corresponds well to areas where the average summer water temperature does not exceed 28 °C (Fig. 9.6). This restricted distribution suggests that the temperature effects on growth and survival of this species limit its distribution, locally around the Japanese Islands.

The culture experiments showed that gamont specimens took at least 15 days at 20 °C to 25 days at 10 °C to grow into adults, and agamont specimens took about 25–40 days. Because of the fairly high growth rate, the life cycle of *P. opercularis* is completed in about 40–65 days. For gamont specimens, small individuals can form plastogamous pairs while growing to the adult stage, and sometimes a small-diameter individual will form a pair with a large-diameter individual. In this case, one full cycle requires a shorter period (Fig. 9.7). Although the culture experiments indicate that the life cycle of this species can be completed in about 55 days, it would take more time under suboptimal environmental conditions. We observed several growth stages coexisting in natural populations (Fig. 9.3), indicating that reproductive timing varies within populations.

Seasonal fluctuations in the abundance of this species reflect its optimal temperature and growth rate (Fig. 9.2). The observed decrease in the number of individuals during August–October is presumably related to higher temperatures during those months. The frequency distribution of shell diameters of this species is uneven and variable through the seasons, suggesting that *P. opercularis* has a short life span (Fig. 9.3).

Effect of temperature on distribution and growth rate of *P. opercularis* were supported by geochemical measurements. Toyofuku et al. (2000) evaluated the Mg/Ca thermometry of this species on the basis of culture experiments and field observations. The Mg/Ca ratio in the high-Mg calcite shells of *P. opercularis* reflects the temperature of growth conditions, and each chamber in an individual specimen reflects the environmental conditions when it calcified. Thus, the life span and the timing of calcification affect the Mg/Ca ratio. In fact, Mg/Ca ratios vary not only between chambers but also, as shown by electron probe microanalyzer (EPMA) observations, within a chamber wall (Toyofuku and Kitazato 2005).

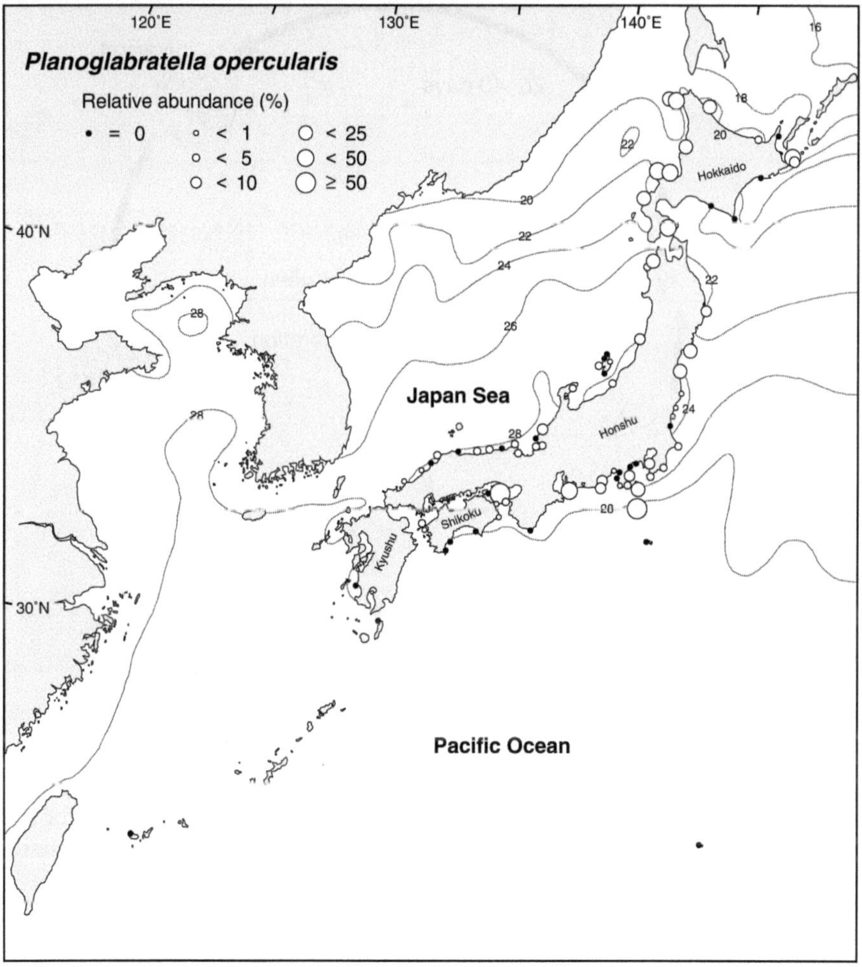

Fig. 9.6 Maximum summer sea surface temperatures (°C), distribution and relative abundances (%) of *P. opercularis* along the Japanese coast

9.3.3 *Determination of Breeding Populations Within Glabratellids and Between Local* P. opercularis *Populations*

Interspecific breeding experiments were carried out using three morphogroups, *G. patelliformis*, *P. nakamurai* and *P. opercularis*, to see if these groups represented interbreeding populations, despite their different phylogenetic relationships (see details in Sect. 9.4) (Fig. 9.4). Gamont individuals from the same parental agamont sometimes formed pairs; however, there was no exchange of gametes. This suggests that autogamy may not occur in glabratellids, even though autogamy has been found

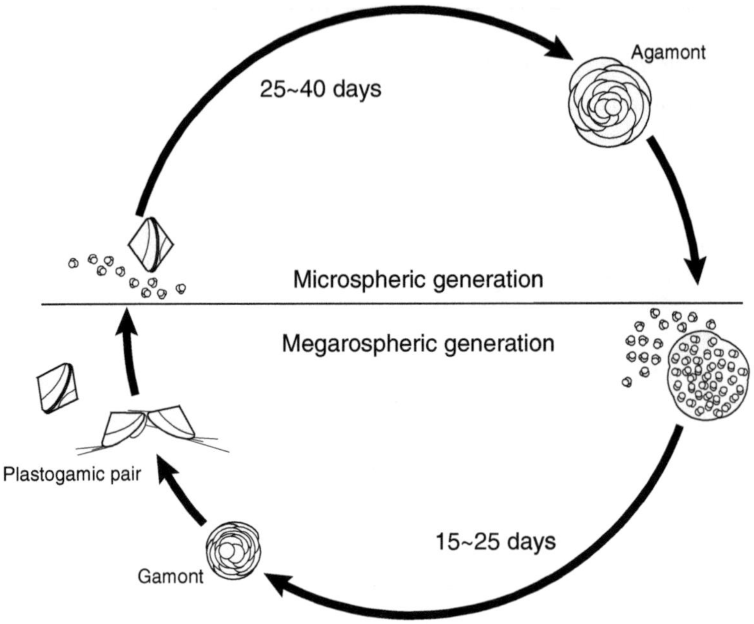

Fig. 9.7 Presumed life cycle of *P. opercularis*, based on culture experiments

among species of the genus *Rotaliella* (Grell 1954, 1957). Individuals from the different morphospecies never reacted to each other, even when their pseudopods were close enough to touch. Individuals from the same morphospecies formed plastogamic pairs, but those from different morphospecies did not make plastogamic pairs (Kitazato et al. 2000).

Intraspecific breeding experiments were carried out using morphological variants of *P. opercularis* (Kitazato et al. 2000). Two morphological variations occur in *P. opercularis*: peripheral spines may be present or absent, and differences in the height:diameter (H:D) ratio, which are apparent as "high" and "low" trochospirals. Individuals both with and without peripheral spines, and with both high and low trochospirals were able to form plastogamic pairs and reproduce during the culture experiment. These results clearly show that all variants belonged to the same population and could interbreed.

Intraspecific (inter-populational) breeding experiments among *P. opercularis* populations from different localities indicated that individuals from geographically distant populations do not form gamontogamous pairs. Gamont specimens from Omaezaki and Shimoda (approximately 100 km distant from Omaezaki along shoreline) formed plastogamic pairs and reproduced. However, individuals from Omaezaki and Sendai (approximately 500 km distant from Omaezaki) or Omaezaki and Echizen-Matsushima (approximately 2,500 km distant from Omaezaki) did not form plastogamic pairs, even though the *P. opercularis* populations from these localities share the same morphological characteristics.

These interbreeding experiments using individuals from widely separated populations show that the interbreeding capabilities of populations are closely related to the geographic distances between them. This suggests that *P. opercularis* constitutes a ring species, i.e., it is composed of a chain of small populations that are reproductively isolated (Kitazato et al. 2000). Our results suggest that specimens collected from locations closer than ~100 km can interbreed, whereas individuals from populations separated by about 500 km cannot interbreed. Note that these observations are based on the results of breeding experiments using specimens maintained in laboratory cultures.

It is noteworthy from a population-genetics perspective that the glabratellids have a life form that crawls on coralline algae and that gamete exchange occurs within a plastogamic pair. As a result, the dispersal potential of both gametes and individuals is most likely low. Glabratellids can therefore become geographically isolated, and this isolation may be affected by the distribution of land and ocean as well as by oceanic currents. This makes *P. opercularis* a useful species for observing genetic differentiation and studying the mechanisms behind the formation of local populations.

9.4 Molecular Phylogenetic Analysis of Glabratellid Species: Morphology, Molecular Phylogeny and Breeding Ability

9.4.1 Molecular Phylogenetic Analysis and Morphological Characteristics Among Japanese Glabratellid Species

Molecular phylogenetic reconstructions based on SSU and LSU rRNA genes correspond well to the morphological characters in glabratellids and planoglabratellids (Tsuchiya et al. 2000). The branching patterns inferred from both LSU and SSU sequences indicate that *Glabratella* species diverged first, followed by *Angulodiscorbis* and both *Planoglabratella* species. *P. opercularis* and *P. nakamurai* form a clade. A comparison of the molecular phylogeny with the results of the interspecific breeding experiments with glabratellids revealed that specimens from the same breeding population show a divergence below 0.8 % in SSU rRNA and below 1.1 % in LSU rRNA. On the other hand, genetic divergences over 0.8 % in SSU and 1.1 % in LSU suggest different breeding populations. These molecular analyses are therefore able to determine relationships at the genus and species levels.

The phylogenetic branching patterns of Glabratellidae show good agreement not only with morphological clusters but also with interspecific breeding capability. The morphospecies that are closely related according to the molecular phylogenetic tree possess the same ventral morphology. The basal species of *Glabratella* and *Angulodiscorbis* have radial ridges made from a straight lined structures of needles; *Planoglabratella* have umbilical bosses (Tsuchiya et al. 2000). The genetic evolution of glabratellids and planoglabratellids is associated with morphological change. A comparison between morphological plasticity during the ontogenetic stages (Kitazato et al. 2000) and molecular phylogeny (Tsuchiya et al. 2000) suggests that these morphological characters are genetically controlled. Breeding populations of planoglabratellid species are characterized by specific combinations of morphological characters.

9.4.2 Genetic Relationships and Morphology Among Planoglabratellid Species

The closely related species *P. opercularis* and *P. nakamurai* are distinguishable on the basis of the ITS of nuclear rRNA gene sequences. Because the ITS sequence is 3–5 times more variable than that of the LSU (Tsuchiya et al. 2003), there are $2.5–4.2 \times 10^{-9}$ substitutions/site/year in the ITS and $0.8–1.7 \times 10^{-9}$ substitutions/site/year in the LSU between *P. opercularis* and *P. nakamurai*. As a result of this rapid evolutionary rate of the ITS, the evolutionary relationships between closely related planoglabratellids can be clearly revealed in detail. At least two genotypes were observed in *P. opercularis*: genotypes A and B. Interestingly, genotype B clusters with *P. nakamurai*, but coiling form of *P. opercularis* genotype B is similar to genotype A that has an evolute coiling form, whereas *P. nakamurai* has an involute coiling form. However, other morphological features of genotype B suggest that it is an intermediate form between genotype A and *P. nakamurai*, with a flattened umbilical side and smooth umbilical bosses, representing a cryptic species within *P. opercularis*, even though both genotype A and B share the same evolute coiling form. The morphospecies that are closely related in the molecular phylogenetic tree possess the same ventral sculpture (Tsuchiya et al. 2003).

For genotype A, genetic distances are greater between specimens from widely separated populations than between those geographically closer together. Sequence variations within local populations of *P. opercularis* genotype A ranged from 0 % to 1.9 % and increased for more distant populations. For example, variation ranged from 0.9 % to 2.4 % between the Omaezaki and Otaru populations (~2,000 km distant), from 3.1 % to 3.7 % between Omaezaki and Echizen-Matsushima (~2,500 km distant), and from 2.4 % to 3.4 % between Otaru and Echizen-Matsushima (~1,500 km distant).

9.4.3 Stable Morphologic Characters

We observed morphologically stable and ecophenotypic variants during the culture experiments and molecular phylogenetic analyses. Trochospiral, spiro-convex, and conical shell share morphological features among glabratellid and planoglabratellid species. The umbilical side is flattened with a slightly depressed umbilicus and is ornamented with rows of pustules that form radially aligned striae. In general, sutures are flush with the surface on the spiral side, but depressed on the umbilical side. Radial striae, which form groove-and-ridge systems, are developed on the ventral face. Radial striae probably have a functionality related to pseudopodial activity during movement or feeding (Kitazato 1992).

Interestingly, the planoglabratellid species display a unique morphologic change within the alternation of generations (Table 9.1). The microspheric generation has a sinistral form, and the megalospheric generation has a dextral form (Fig. 9.8) having same nucleotide sequence (Tsuchiya et al. 2000, 2003). We observed sinistral adult agamonts producing dextral clone juveniles, and two dextral parental cells forming a plastogamic pair and producing sinistral agamont juveniles. One hundred percent of sexually reproduced offspring of agamonts showed sinistral coiling; in contrast,

Table 9.1 Differences in coiling direction between (a) sexually reproduced offspring (agamont) and (b) asexually produced offspring (gamont) at different culture temperatures

	10 °C	15 °C	20 °C
(a)			
Dextral	0	0	0
Sinistral	12	52	39
Coiling direction of parents	D-D	D-D	D-D
Number of pairs observed	1	3	2
(b)			
Dextral	122	77	684
Sinistral	3[a]	0	8[a]
Coiling direction of parent	S	S	S
Number of adults observed	2	2	12

[a]Juveniles were in close proximity and two or more would convolute and make new chambers with abnormal coiling. *D* dextral coiling direction, *S* sinistral coiling direction

although almost all asexually produced offspring of gamonts showed dextral coiling, a small percentage (less than 1.2 %) coiled in the opposite direction. These latter cases involved juveniles in close proximity where two or more individuals convoluted when forming new chambers resulting in abnormal coiling, or became connected to each other in rare cases. However, because almost 99 % of asexually reproduced individuals show dextral coiling, the coiling direction can be considered a stable character during the ontogenetic stages (Kitazato et al. 2000). Myers (1938) described a similar alternation of coiling direction for *G. sulcata* in a figure (Plate IB in Myers 1938); therefore, this morphological feature is presumably shared among glabratellid species. This unique morphological differentiation could be useful in discovering genes related to morphogenesis and also to calcification and certain proteins.

The proloculus diameter does not change between generations of *P. opercularis*. We observed the proloculus diameter just after reproduction; agamont specimens from two parental plastogamic pairs had an average diameter of 60.9 μm, and 23 gamont specimens produced from a single agamont specimen had an average diameter of 61.0 μm (Table 9.2). Even though this sample size is too small for a statistical comparison, it appears that the proloculus diameter is stable between samples from the same generational line. It is possible that the proloculus diameter changes depending on the volume of the parental cell or the mating process. In fact, the average proloculus diameter from two sets of sexually produced offspring differed by 9.3 μm. Erskian and Lipps (1987) suggested that coiling direction in natural populations of *Glabratella ornatissima* is based on proloculus diameter. This change in size of proloculus diameter from microspheric to megalospheric with the alternation of generations is characteristic in general, especially in the larger foraminiferal species; *Archaias angulatus*, *Calcarina gaudichaudii*, *Heterostegina depressa*, and others (e.g., Ross 1972; Röttger 1974; Röttger et al. 1990; Hallock et al. 1986). The coiling pattern in *P. opercularis* changes with each generation, the agamont having a low trochospiral and the gamont having a high trochospiral shell. As a result, the posterior whorl covers chambers that previously appeared at the apex of the shell on the surface of the individual, and the large area of proloculus exposed at the surface at adult stage in the gamont rather than the agamont.

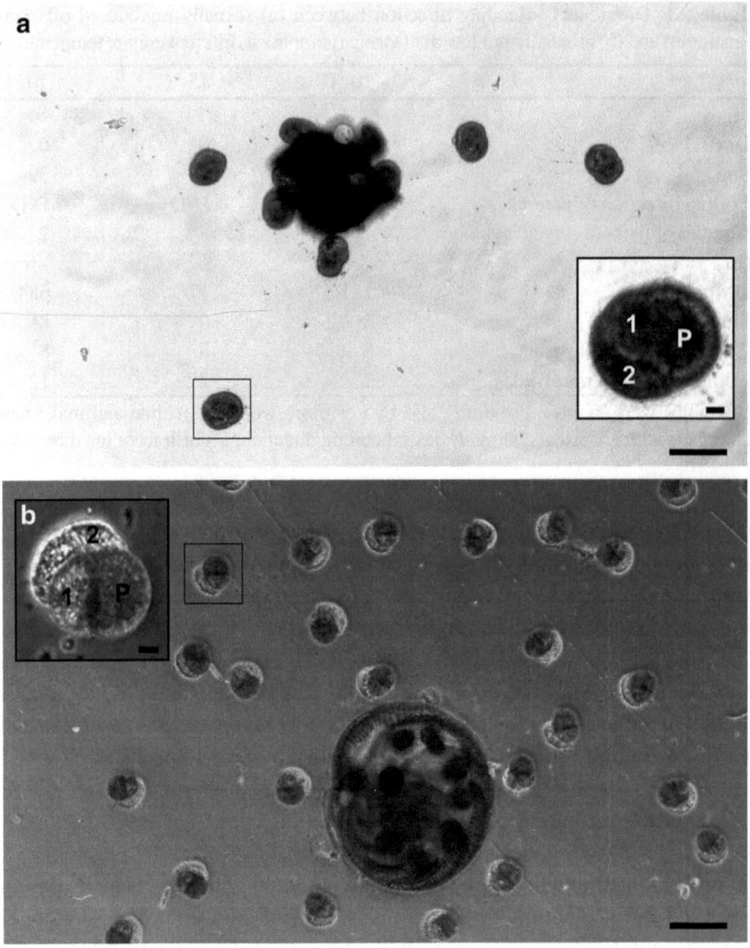

Fig. 9.8 Reflected light micrograph of coiling direction of juvenile *P. opercularis*. The third chamber appears just after reproduction. Both sexually (**a**) and asexually (**b**) reproduced offspring have a similar-sized proloculus. Scale bar, 100 μm. *Inset* indicates magnified juvenile specimen. Scale bar, 10 μm. Proloculus (*P*), first chamber (*1*), second chamber (*2*)

Table 9.2 Differences in proloculus diameter between sexually reproduced offspring (agamont) and asexually reproduced offspring (gamont)

	Sexually reproduced offspring (agamont)			Asexually reproduced offspring (gamont)
	Total	[Pair #1]	[Pair #2]	Total
Average proloculus diameter (μm)	60.9	(56.3)	(65.6)	61.0
Standard deviation (μm)	6.4	(4.4)	(4.8)	3.8
Number of observed specimens	16	(8)	(8)	23
Number of parental pairs	2	(1)	(1)	1

Fig. 9.9 Scanning electron micrograph showing ecophenotypical morphologic variations in *P. nakamurai* specimens (1–6) from the same Petri dish. Dorsal view (*a*), side view (*b*) and ventral view (*c*). Scale bar, 100 μm

9.4.4 Ecophenotypic Morphological Variations

Several morphological characters such as the presence of peripheral spines and the H:D ratio vary with growth stage and also with the ambient environment (Kitazato et al. 2000). The peripheral spines of *P. nakamurai*, which is genetically closely related to *P. opercularis*, disappear at later growth stages, becoming covered by non-spinose chambers (Kitazato et al. 2000). The involutely coiled whorl in *P. nakamurai* sometimes becomes evolute at the mature stage. As with *P. nakamurai*, there are both spinose and non-spinose individuals of *P. opercularis* (Kitazato et al. 2000). However, peripheral spines of *P. opercularis* morphogroup appear at all ontogenetic stages. This phenomenon was not observed in *P. nakamurai*. In addition, this morphological variation in *P. opercularis* is observed at all localities around the Japanese Islands. The presence or absence of spines and the height of the shell are ecophenotypic characters, with the development of spines controlled genetically in this species. The length of the spines and the height of the shell are affected by the ambient environment. In fact, in *P. nakamurai* the length and position of spines of individual specimens can change within the same Petri dish (Fig. 9.9).

Compared to other benthic foraminifera, we can clarify ecophenotypic variants and stable morphological characters within the molecular phylogenetic lineage not only as in planoglabratellid species, but also other rotallids. In *Bulimina* species, including the direction, placement and shape of spines, the angle of undercutting of the chamber periphery, and the roundness of the chambers, are stable among specimens of each clade. In contrast, the length and density of spines and chamber size are variable within each clade of *Bulimina* species (Tsuchiya et al. 2008). In the other case, the development of pore morphologies depends on ambient environmental conditions; in *Virgulinella fragilis*, small pores develop on the surface of the shell in well-oxygenated environments, whereas elongated and connected pores develop in dysoxic environments (Tsuchiya et al. 2009). *Ammonia beccarii* develops significantly larger pores under culture conditions with low dissolved oxygen concentrations than under well-oxygenated conditions (Moodley and Hess 1992). To compare molecular phylogeny and morphology, we investigated morphological differences within populations classified as genetically the same, or within the same biological species (i.e., breeding populations), as well as ontogenetic morphological differences between juveniles and adults, and within clone cultures.

9.5 Genetic Diversification of *P. opercularis*

9.5.1 Genetic Relationships Among Local Populations of P. opercularis *and Their Relatives*

To test whether non-interbreeding populations of *P. opercularis* can be distinguished by genetic differences, and how local populations of this species developed; we examined 36 specimens of this species from 16 locations. To determine genetic differences, we used the internal transcribed spacers (ITS) of ribosomal DNA sequences, especially in ITS 1 region, because it evolves more rapidly than LSU rRNA genes used in previous studies of Glabratellidae (Tsuchiya et al. 2000) and other benthic foraminifera (e.g., Holzmann and Pawlowski 2000). DNA extractions were carried out by using methods described by Holzmann and Pawlowski (1996), and Pawlowski (2000), and PCR conditions and primer sequences were described in Pawlowski et al. (1996) and Tsuchiya et al. (2003). Sequence alignment was carried out with Clustal W (Thompson et al. 1994), and sequence editing were carried out with Se-Al (Rambaut 1996). Molecular phylogenetic reconstructions were made with the following methods: Kimura 2-parameter method for calculating genetic distances (Kimura 1980), Saitou and Nei (1987) for Neighbor-joining method, and Guindon et al. (2010) for Maximum likelihood method.

ITS is composed of three regions: ITS 1, the 5.8S rRNA gene, and ITS 2 (Fig. 9.10). To identify the position of each region in *P. opercularis*, the sequences obtained were aligned with those of other eukaryotes. According to the alignment, the region between bp positions 357 and 573 corresponds to the 5.8S rRNA gene. Part of this region, from positions 370–436 bp (dark gray area in Fig. 9.10a) was conserved among most of the eukaryotes studied. The ITS 1 region spans a fragment of 205 nucleotides between bp positions 152 and 356 in our alignment, and the ITS 2

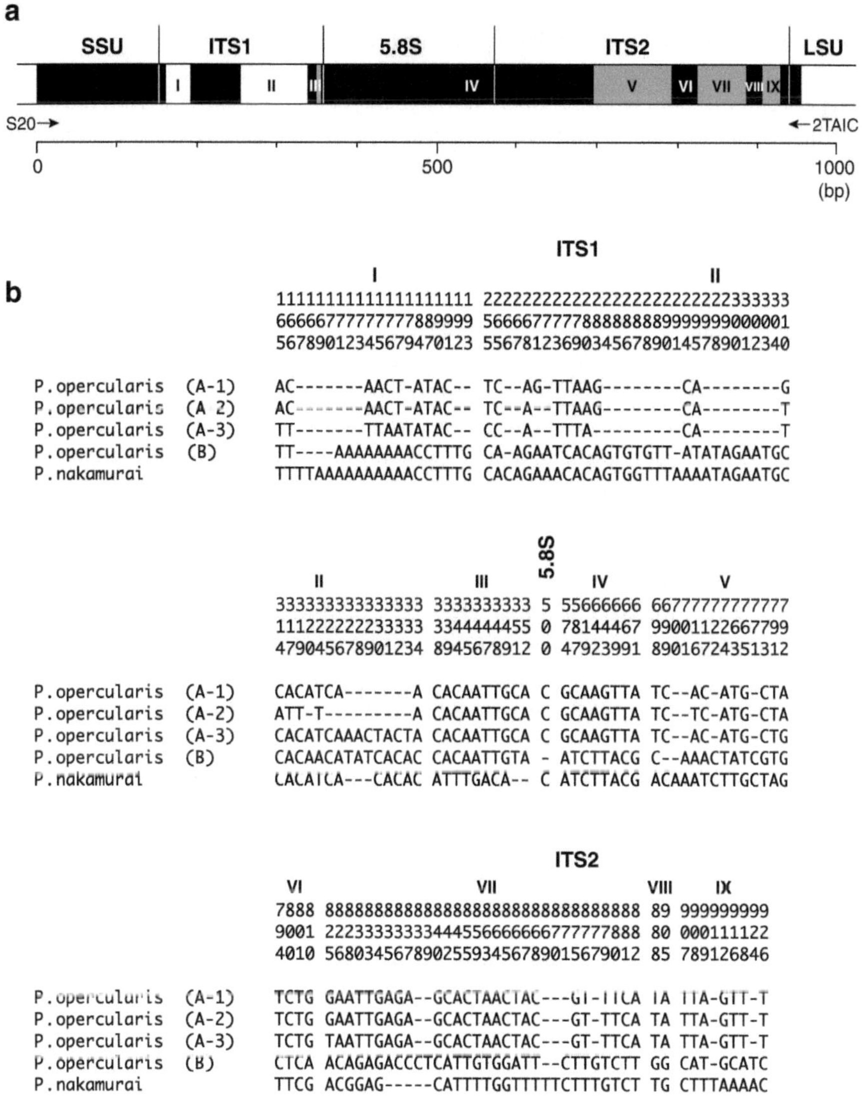

Fig. 9.10 Diagram of the ITS region of nuclear rDNA and variable sites in each segment. Positions are indicated starting from the s20 primer binding site in SSU rDNA to the 2TAIC primer binding site at the 5′ terminal end of the LSU rDNA sequence (**a**) and the nucleotide positions (**b**). The regions conserved among the two genotypes of *P. operuclaris* and *P. nakamurai* (*solid black*); the variable regions (*white*); and the regions conserved among local population of *P. opercularis* genotype A-1, A-2 and A-3 (*light gray*) are shown. The 3′-end of the 5.8S region, which is conserved in most eukaryotes, is indicated by *dark gray* in (**a**). Note that each sequence type is shown as a consensus sequence

region extends from bp positions 574–939 (365 nucleotides in length). The 5.8S rRNA gene and adjacent segments of the flanking ITS 1 and ITS 2 regions (from position 338 to 696) are conserved in all glabratellid species (black areas in Fig. 9.10a).

Small conserved fragments are found also in ITS 1 and ITS 2, but the majority of these regions were variable in the species examined.

Two distinct sequence types have been distinguished within *P. opercularis* (genotypes A and B), and within *P. opercularis* genotype A, there are at least three major distinguishable sequence types (A-1, A-2, and A-3) (Fig. 9.10b). Most of the variations within genotype A occur in the ITS 1 region, particularly in region II of ITS 1. There are only three sites in the ITS 2 region where the sequences of genotype A are variable. Thus, ITS 1 can be used to distinguish differences among local populations. Genetic variations within each clade are relatively small, ranging from 0.0 % to 1.2 % (A-1), 0.0 % to 1.2 % (A-2), and 0.6 % to 2.4 % (A-3) (Tsuchiya et al. 2003). The sequence of *P. opercularis* genotype B resembles that of *P. nakamurai* in regions I and II of ITS 1 and region IV of ITS 2, but it is similar to *P. opercularis* type A in region III of ITS 1 (Fig. 9.10b). Genotype B has not only intermediate nucleotide sequences but also morphological characters that fall between those of genotype A of *P. opercularis* and the *P. nakamurai* genotype. Detailed characteristics of both genotypes have been presented elsewhere (Tsuchiya et al. 2003).

The phylogenetic relationships among *P. opercularis*, *P. nakamurai* and *A. quadrangularis*, inferred from ITS 1 rDNA sequences, are illustrated in Fig. 9.11. At least three distinct clades (A-1, A-2 and A-3) were recognized within *P. opercularis* genotype A, and two minor genotypes ("Novel" genotypes in Fig. 9.11) were distinguished as sister phylotypes of the genotype A and *P. nakamurai*/genotype B clusters. The clades are supported by relatively high bootstrap values of 97 % by the maximum likelihood method (ML) and 85 % by the neighbor joining method (NJ) for the A-1 clade, and 85 % (ML) and 91 % (NJ) for the A-2 clade. In both analyses, clades A-1 and A-2 group together with high bootstrap support, 86 % by ML and 89 % by NJ, whereas clade A-3 has only slight bootstrap support by ML analysis (27 %) but relatively high bootstrap support by NJ analysis (61 %). This suggests that there must be sequence differences in A-3; this genotype shows a wider distribution than genotypes A-1 and A-2. In contrast, the three clades cannot be distinguished by analysis of the more conserved ITS 2 region (data not shown). The divergence of ITS 1 sequences among the three clades of *P. opercularis* genotype A may be as much as 8.1 %. The highest genetic distances are observed between clade A-3 and the other clades. The distances range from 5.5 % to 7.3 % between A-1 and A-3 and from 6.2 % to 8.1 % between A-2 and A-3, whereas clades A-1 and A-2 differ by 2.5 % to 3.7 %. The divergences seem to increase with the geographic distance separating local populations, but this is not always the case. The divergence between ITS 2 sequences is much smaller; however, clade A-3 have slightly different nucleotide sequences from both A-1 and A-2 clades as in the case of ITS 1.

9.5.2 *Geographic Distribution and Natural Abundances of* **P. opercularis** *Type A Populations*

The three clades within *P. opercularis* genotype A distinguished by analysis of ITS 1 sequences can be considered to represent three populations of this species along the

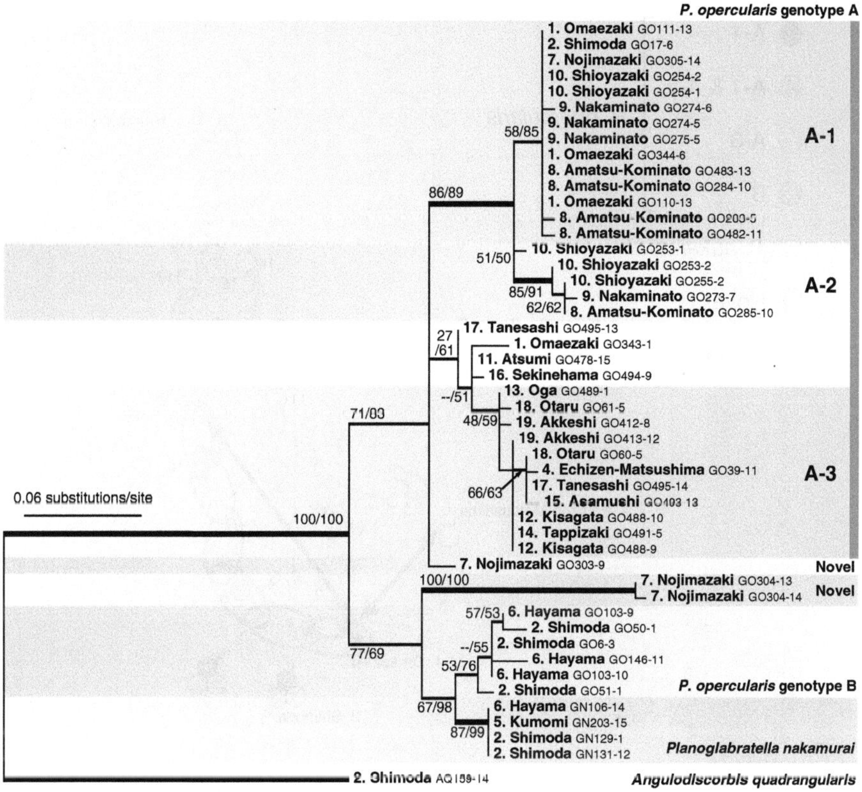

Fig. 9.11 Phylogenetic relationships between *P. opercularis* and *P. nakamurai* inferred by maximum likelihood (ML) and neighbor joining (NJ) methods based on ITS 1 sequences. Three genetically distinct populations of *P. opercularis* are designated as A-1, A-2 and A-3. *A. quadrangularis* was used as the outgroup species. Bootstrap values greater than 50 % are presented above the branches. Each sequence is designated by the name of the collection locality and specimen number

Japanese Islands. Although there are no morphological differences between these populations, each shows a different distribution pattern. The specimens belonging to population A-3 were found primarily along the Japan Sea coast of Honshu and around Hokkaido Island. Populations A-1 and A-2 were limited to the Pacific side of Honshu, with A-2 present only in the northern part of the island (Fig. 9.12).

Our data strongly support the ring species hypothesis proposed by Kitazato et al. (2000). The genetic differences observed between distant populations of *P. opercularis* are congruent with their reproductive isolation as suggested by breeding experiments. As described in Sect. 9.3.3, we found incompatibilities in breeding behavior in inter-populational breeding experiments using local populations of *P. opercularis*, showing that members of geographically distant populations do not form plastogamic pairs or reproduce sexually, represent genetic variation between locations with genetic clines in relation to geographic distances. In particular, there were

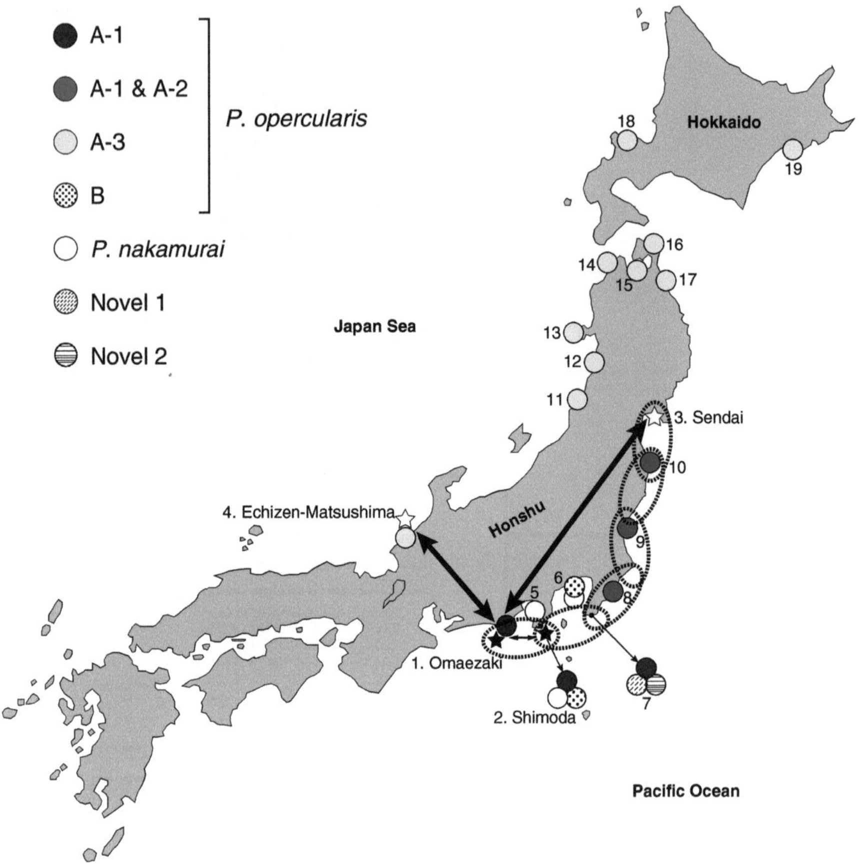

Fig. 9.12 Distribution of ITS 1 genotypes and local populations of *P. opercularis* genotype A. Breeding experiments used specimens from four localities (Omaezaki, Shimoda, Sendai and Echizen-Matsushima), and were conducted against specimens from Omaezaki (*solid and open stars*). *Solid stars* (and *a solid double-headed arrow*) indicate breeding ability between locations, and *open stars* (and *dashed double-headed arrows*) indicate unsuccessful breeding with Omaezaki specimens. *Dashed circles* indicate the putative ring-like structure of the breeding population inferred from breeding experiments. The location numbers correspond to those represented in the phylogenetic tree in Fig. 9.11. Numbers indicate sampling locality: *1*. Omaezaki, *2*. Shimoda, *3*. Sendai, *4*. Echizen-Matsushima, *5*. Kumomi, *6*. Hayama, *7*. Nojimazaki, *8*. Amatsu-Kominato, *9*. Nakaminato, *10*. Shioyazaki, *11*. Atsumi, *12*. Kisagata, *13*. Oga, *14*. Tappizaki, *15*. Asamushi, *16*. Sekinehama, *17*. Tanesashi, *18*. Otaru, *19*. Akkeshi

large genetic differences between the Japan Sea population (A-3) and the Pacific populations (A-1 and A-2).

The populational structure of *P. opercularis* suggests a ring species, in which a chain of small local populations that can interbreed only with genetically homogeneous neighboring populations. One reason for geographic isolation in this species might be the low dispersal ability of gametes. Like other glabratellids, *P. opercularis* reproduces sexually by gamontogamy; that is, its gametes are released and form

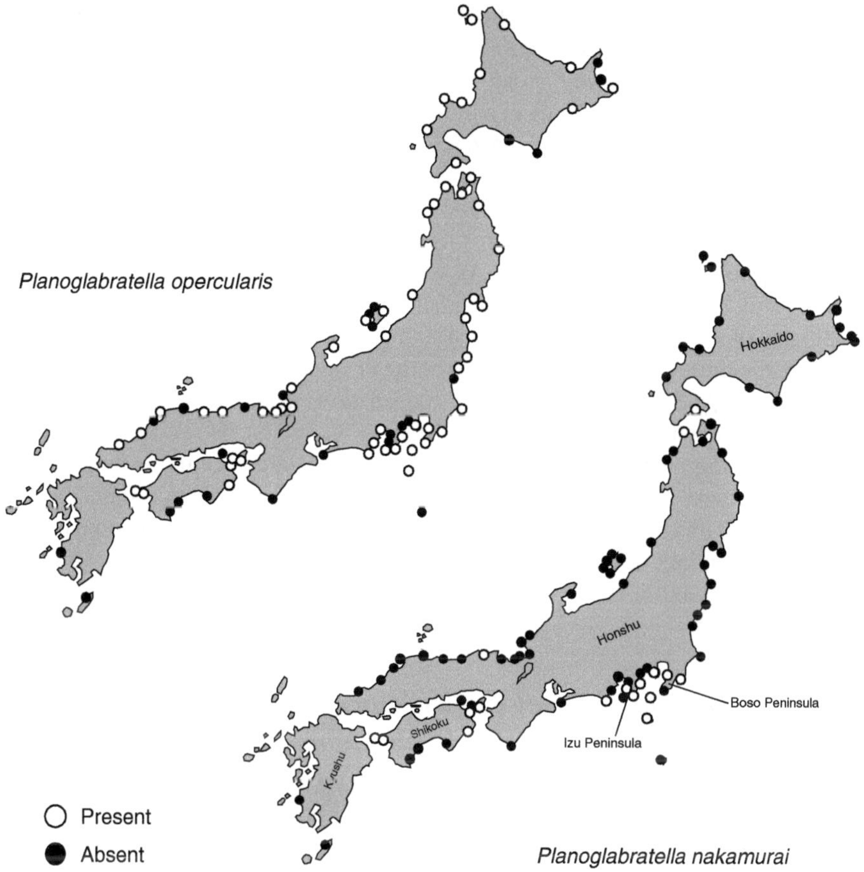

Fig. 9.13 Distribution of morphospecies *P. opercularis* and *P. nakamurai* during 1985–1986

zygotes within the enclosed space formed by the shells of a gamontogamous pair of parental cells (Myers 1940; Grell 1979). The glabratellids also show relatively low motility as adults because of their mode of life; they live on coralline algae, crawling on thalli (Kitazato 1988, 1992, 1994).

Therefore, *P. opercularis* was easy to isolate geographically, as we expected. Our results suggest a good correlation between genetic populations and breeding populations. The genetic cline of *P. opercularis* genotype A can be explained by the low dispersal ability of gametes resulting from their plastogamic sexual reproduction. In the case of *P. opercularis* genotype A (indicated by a monophyletic relationships among LSU or SSU analyses), we can discriminate local populations through ITS 1 sequences. Sequence divergences below 2 % indicate that individuals are from the same breeding population, whereas sequence divergences over 5 % suggest sexually isolated local populations. A sequence divergence of approximately 3 % suggests that weak isolation has occurred between local populations.

The geographic distribution patterns of *P. opercularis* and *P. nakamurai* are different, and both morphospecies coexist in some local populations (Fig. 9.13).

Interestingly, the novel genotype from the ITS 1 tree (Fig. 9.11) containing high genetic variability was found in specimens from Nojimazaki. Nojimazaki is an area of overlap of both *P. nakamurai* and the closely-related A-1 and A-2 genotypes. Genotype B was also found only in a limited area where the distributions of *P. opercularis* and *P. nakamurai* overlapped. It is possible that several genotypes were produced as interspecific or inter-populational hybrids, creating a hybrid zone. Pillet et al. (2012) proposed interspecific hybridization in *Elphidium macellum*, and that expansion-segment polymorphism originated from hybridization and is associated with morphological differentiation. As we described in Sect. 9.4.2, *P. opercularis* genotype B shows morphological characteristics intermediate between the morphologies of *P. opercularis* genotype A and *P. nakamurai* (Tsuchiya et al. 2003). At the same time, the primary structures of the ITS nucleotide sequences of *P. opercularis* genotype B share features of both *P. opercularis* genotype A and *P. nakamurai* nucleotide sequences (Tsuchiya et al. 2003) (Fig. 9.10b). However, we did not observe the formation of plastogamic pairs between interspecific or interpopulational specimens during our breeding experiments. It is possible that some mechanism supports this interspecific relationship; the possible occurrence of interspecific hybridization might be clarified by the analysis of organelle-encoded genes such as those in mitochondria.

9.5.3 *Formation of Local* **P. opercularis** *Type A Populations and Paleoceanographic Implications*

We focused on genotype A, the distribution of the three local populations may be related to the geologic history of the Japanese Islands, in particular to paleoceanographic changes that occurred during the late Cenozoic (Fig. 9.14a-d). The Japanese Islands appear to have acted as a barrier, preventing gene flow between Japan Sea and Pacific Ocean populations, and recent distribution of genetic populations may occur with ocean current, passively dispersed by drifting seaweeds (Spindler 1980) or propagules (Alve and Goldstein 2002, 2003). The phylogeneitc tree was reconstructed with selected nucleotide sequences among planoglabratellid species and local populations of *P. opercularis* genotype A based on maximum likelihood method (Fig. 9.14e). We also calculated the divergence time based on molecular dating using a single calibration point as a fixed age (divergence of *P. opercularis*) by using a phylogenetic program MEGA5 (Tamura et al. 2011) and found that they are consistent with the molecular, fossil record and paleoceanographic land-ocean distribution, although these assumptions have a margin of error, because of difficulty in preservation of rocky shore foraminifers as fossil record.

Low dispersal ability may not be the only reason for the differentiation of *P. opercularis* populations. Their origins can also be explained by paleoceanographic events related to the history of the Japanese Islands. According to recent studies (Otofuji and Matsuda 1984; Hirooka 1988), the continental margin crust that forms the foundation of the Japanese Islands separated from the Eurasian continent around 17–15 Ma, producing a marginal sea. During 15–10 Ma, small islands in the northeastern region of the present-day Japanese Islands were surrounded by relatively shallow water. Diverse benthic molluscan fauna were adapted to this shallow-water environment (Chinzei 1991). An upheaval between 10 and 6 Ma formed the northeastern Japanese Islands (Fig. 9.14a). At around 4 Ma, the land area located approximately at present-day position of the Japanese Islands (Fig. 9.14b) (Chinzei 1991; Ogasawara 1994), both the oceanography and biogeography around the Islands changed drastically. Benthic molluscs associated with cold-water environments were separated into the Omma-Manganji Fauna in the Japan Sea and the Tatsunokuchi Fauna along Pacific Coast (Chinzei 1991). After the formation of the Japanese Islands, the Japan Sea was connected to the open ocean by three shallow, narrow straits: the Tsushima (~100 m water depth), Soya (~60 m water depth), and Tsugaru straits (~450 m water depth). Then, during the last glacial period (Fig. 9.14c) (Oba et al. 1991), the Japan Sea experienced drastic environmental changes because as a result of sea-level regression it became geographically isolated from surrounding oceans. This geographic isolation of the Japan Sea allowed the divergence of genetically different populations.

The oldest fossil records of *P. opercularis* around the Japanese Islands were reported from the strata dating to 17–15 Ma in northern Japan (Fujita and Ito 1957; Matoba et al. 1990). No older fossil records of this species from around the Japanese Islands have been reported. The paleogeographic reconstruction suggests that the ancestral *P. opercularis* population may have been distributed in marginal seas around this time, when the Japanese Islands may not have served as a barrier to genetic mixing. The key geological event for the origin of different *P. opercularis* populations was probably the formation of the Japanese Islands by an uplift event around 4 Ma. Almost all present-day coastlines of the Islands developed at around this time period. Since then, the Islands have acted as a topographic barrier between the Japan Sea and the Pacific Ocean. The Japan Sea population (A-3) probably became isolated from the Pacific populations at about that time. If it is assumed that the divergence of *P. opercularis* in the Japanese Islands occurred at 17–15 Ma, then the branching of the A-3 population can be dated to 8.1 Ma (Fig. 9.14e). This chronology supports the hypothesis that the timing of branching events coincides with the development of the Japanese Islands.

The divergence of the A-1 and A-2 populations probably occurred more recently. According to molecular data, their estimated time of divergence averages 2.9 Ma (Fig. 9.14e). This divergence may be related to geographical separation following sea-level changes, or to the collision between the Izu-Ogasawara Arc and Honshu Arc (Kaizuka 1984; Nakamura et al. 1984), proposed to have occurred around 1 Ma (Kitazato 1997). The isolation of the A-1 and A-2 populations is related to this tectonic event or to changes in oceanic currents. Northern Izu was located close to the Honshu Arc around this time. It is possible that the northern part of the Izu-Ogasawara Arc acted as a geographic barrier. However, as the phylogenetic analysis of the ITS 2 region shows, genetic mixing between both populations cannot be excluded, and they coexist in three localities (Shioyazaki, Nakaminato, and Amatsu-Kominato). The phylogenetic tree of ITS 2 also shows unresolved relationships

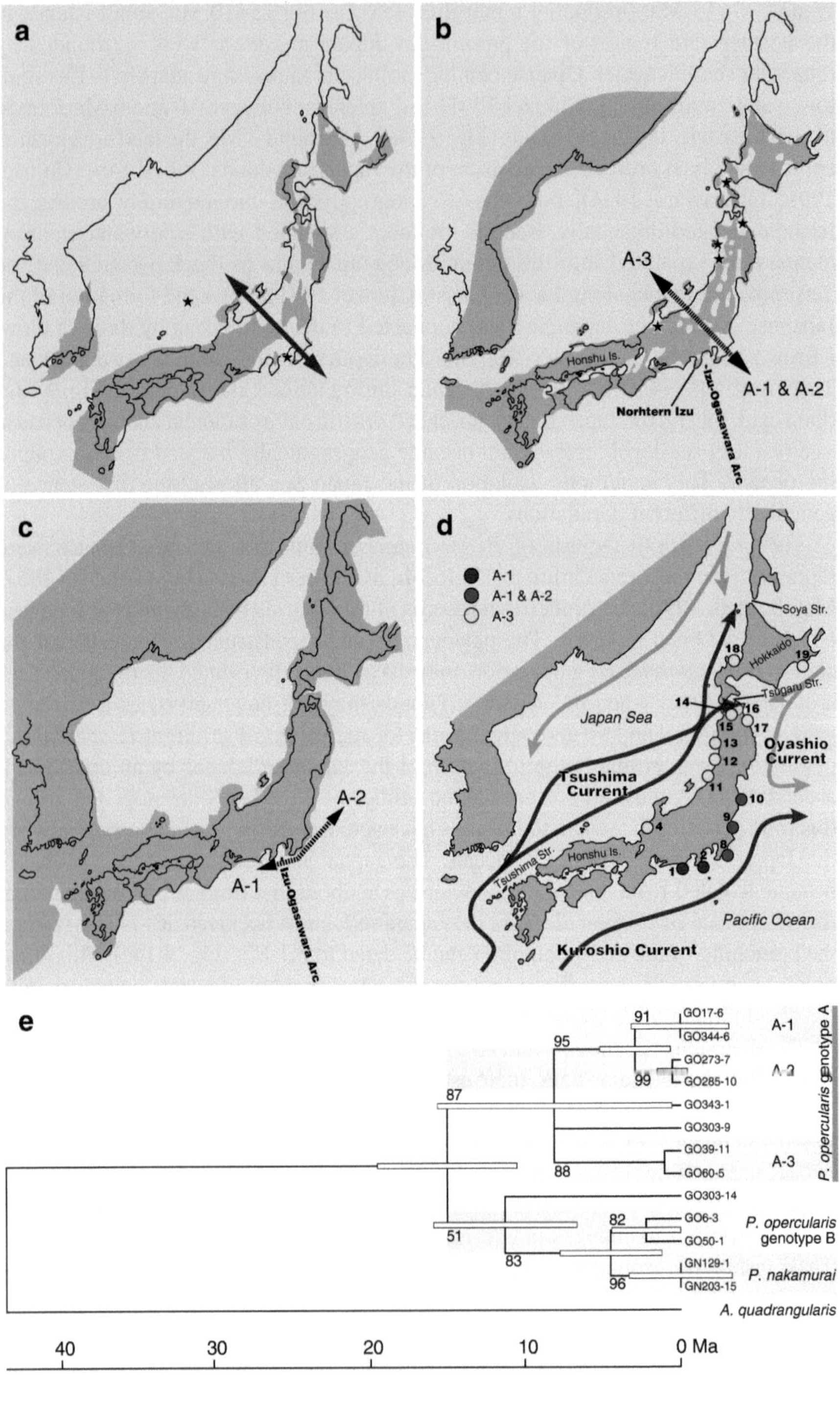
a
b
A-3
A-1 & A-2
Honshu Is.
Norhtern Izu
Izu-Ogasawara Arc
c
A-2
A-1
Izu-Ogasawara Arc
d
A-1
A-1 & A-2
A-3
Soya Str.
Hokkaido
Tsugaru Str.
Japan Sea
Oyashio Current
Tsushima Current
Honshu Is.
Tsushima Str.
Pacific Ocean
Kuroshio Current
e
GO17-6
GO344-6
A-1
GO273-7
GO285-10
A-2
GO343-1
GO303-9
GO39-11
GO60-5
A-3
P. opercularis genotype A
GO303-14
GO6-3
GO50-1
P. opercularis genotype B
GN129-1
GN203-15
P. nakamurai
A. quadrangularis
40
30
20
10
0 Ma

between the A-1 and A-2 populations (data not shown). These relationships suggest that genetic mixing between these populations is probably occurred until recently. In addition, the present-day distributions of *P. opercularis* populations have probably been affected by migration. The A-1 population, originally found in central Japan, is probably migrating toward the northeast with the Kuroshio Current. In the same manner, the A-3 population has spread from Echizen-Matsushima to Hokkaido Island with the Tsushima Current (Fig. 9.14d).

The A-3 population is also found at Akkeshi. However, this could be evidence of a relatively recent migration. The distribution of the A-3 population might be related to the northward shift of warm-water molluscs at 6,000 years B.P. along with changes in oceanic currents. In fact, the Japan Sea shallow-water molluscan fauna migrated toward the northern, Pacific Ocean side of Hokkaido Island around 6,000 years B.P., under the strong influence of the Tsushima Warm Current (Matsushima 1984). Therefore, it is possible that the A-3 population was likewise able to spread to the Pacific side of Hokkaido Island. During the last glacial maximum (LGM), around 18 ka, the Japan Sea was separated from the surrounding seas; the only input was low-salinity surface water (27 psu) introduced via the Tsushima Strait. During this period, vertical mixing was reduced, and the deep water became stagnant. Our interpretation of events suggests that the A-3 population in the Japan Sea survived through this period. Our culture experiments showed that *P. opercularis* can grow and reproduce under low-salinity conditions, even though its growth rate is reduced, and the salinity in the intertidal zone of rocky shore environments (i.e., tide pools) changes from low to high salinity; thus, the A-3 population probably survived under these conditions during the LGM.

9.6 Summary and Future Perspectives

This investigation into the species concept and its relationship to actual evolutionary history has been highlighted by multidisciplinary research. We propose an explanation for the diversification of local populations of *P. opercularis* and their inter- and

Fig. 9.14 Paleoceanographic interpretations of both genetic differentiation and dispersal of local populations among *P. opercularis* genotypes A. Paleogeographic maps are modified from Chinzei (1986, 1991) and Ogasawara (1994). Maps show land areas (*gray*) from the Middle Miocene to Recent. (**a**) 10–6 Ma, (**b**) 5–1.5 Ma, (**c**) 18 ka, (**d**) recent. *Double-headed arrow* (*dark gray*) indicates bidirectional spreading of the *P. opercularis* populations (**a**). *Double-headed dashed arrow* indicates the separation of Japan Sea and Pacific populations (**b**). *Double-headed dashed arrow* in (**c**) indicates the separation of A-1 and A-2 populations. *Black* and *light gray arrows* indicate Kuroshio warm current and Oyashio cold current, respectively (**d**); Present-day distributions of each genotype appear to be associated with these currents. *Solid stars* indicate locations where fossil *P. opercularis* specimens were found, as reported by Asano (1951), Higuchi (1954), Fujita and Ito (1957), Hasegawa (1979), Matoba et al. (1990), Nomura (1992), and Nemoto and Yoshimoto (2001). Estimation of divergence time of selected nucleotide sequences among planoglabratellid species and local populations of *P. opercularis* genotype A (**e**) intraspecific relationships, based mainly on culture and breeding experiments. The relationship and meaning of genetic lineages and genetic distances have been clearly defined through multidisciplinary research as above. Molecular phylogenetic analysis was conducted on specimens of *P. opercularis* from the Japanese Islands by using the internal transcribed spacer of nuclear ribosomal DNA (ITS rDNA). Based on differences in the ITS 1 region, three distinct populations, A-1, A-2 and A-3, were recognized to have genotype A, one of two *P. opercularis* phylotypes. Each population shows a different distribution pattern. The A-3 population is found in the Japan Sea and around Hokkaido Island, whereas populations A-1 and A-2 are present along the Pacific coast of Honshu Island. The distribution of the three populations can be related to the geologic history of the Japanese Islands, in particular to paleoceanographic changes that occurred during the late Cenozoic. The Japanese Islands seem to act as a barrier, which prevents gene flow between Japan Sea and Pacific populations. In addition, the genetic isolation of local Japan Sea populations can be attributed to the low dispersal abilities of *P. opercularis* resulting from their mode of life and gamontogamous reproductive behavior.

In some cases, morphospecies of benthic foraminifers are distributed worldwide as cosmopolitan species. Although molecular phylogenetic analyses have revealed cryptic speciation and defined trans-oceanic genetic exchange and dispersal (e.g., Ertan et al. 2004; Schweizer et al. 2005; Grimm et al. 2007; Pawlowski and Holzmann 2008), previous studies did not clarify the mechanisms of foraminiferal speciation. As glabratellids become geographically isolated, genetic differences accumulate in each local population and sexual isolation may occur (Tsuchiya et al. 2000, 2003; this article).

The mechanisms of genetic diversification differ from species to species. Widely distributed but disjunct populations of *V. fragilis*, which live in oxygen-depleted environments, are genetically the same, although it is difficult to imagine any environmental connectivity between these disjunct populations in terms of current or water mass conditions (Tsuchiya et al. 2009). The genetic connectivity between species of deep-sea epifaunal foraminifera from the Southern Ocean and the Arctic is high, but that between infaunal foraminifera is low (Pawlowski et al. 2007). Interspecific hybridization in *E. macellum* offers a novel perspective on foraminiferal speciation (Pillet et al. 2012). At the other extreme, genetic divergence within planktonic foraminifers is affected by currents, water mass conditions, and glacial–interglacial cycles, and thus speciation can occur by both vicariance (Darling et al. 2004) and dispersal (Ujiié et al. 2012).

Evolutionary patterns in the formation of local populations are different for each species. Ecological characteristics of foraminiferal species, such as their habitat and mobility (including passive migration by propagules or gamete transportation), affect their genetic connectivity, and we have already discussed vicariance and dispersal to some degree. In the open ocean, there are no clear barriers; do the same mechanisms operate? In the case of pelagic fishes, current and water-mass boundaries affect the direction of genetic connectivity (Miya and Nishida 1997). Planktonic foraminifers, however, sometimes migrate across water-mass boundaries, and they show daily vertical migration or migrate with the lunar cycle. To reveal the diversification mechanisms in the foraminifera, it is necessary to understand not only genetics but also adaptations such as food preferences and nutritional requirements that make ecological contributions to speciation. To this end, there is an increased need for culture experiments in the study of foraminifera in combination with multidisciplinary approaches such as in situ (tracer) experiments, and research into genomics and metabolism.

References

Aizawa M (1987) The seasonal changes in benthic foraminifera at the rocky shore. Undergraduate thesis, Institute of Geosciences, Shizuoka University, p 40 (in Japanese with English abstract)

Alve E, Goldstein ST (2002) Resting stage in benthic foraminiferal propagules: a key feature for dispersal? Evidence from two shallow-water species. J Micropalaeontol 21:95–96

Alve E, Goldstein ST (2003) Propagule transport as a key method of dispersal in benthic foraminifera (Protista). Limnol Oceanogr 48:2163–2170

Asano K (1951) Illustrated catalogue of Japanese Tertiary smaller foraminifera, part 14. Hosokawa Printing Co, Tokyo, pp 1–14

Chinzei K (1986) Faunal succession and geographic distribution of Neogene molluscan faunas in Japan. In: Kotaka T (ed) Japanese Cenozoic mulluscs-their origin and migration. Paleontological Society of Japan, Special papers, no. 29, pp 17–32

Chinzei K (1991) Late Cenozoic zoogeography of the Sea of Japan area. Episodes 14:231–235

Darling KF, Wade CM (2008) The genetic diversity of planktic foraminifera and the global distribution of ribosomal RNA genotypes. Mar Micropaleontol 67:216–238

Darling KF, Kucera M, Pudsey CJ, Wade CM (2004) Molecular evidence links cryptic diversification in polar planktonic protists to Quaternary climate dynamics. Proc Natl Acad Sci USA 101:7657–7662. doi:10.1073/pnas.0402401101

Erskian MG, Lipps JH (1987) Population dynamics of the foraminiferan *Glabratella ornatissima* (Cushman) in northern California. J Foramin Res 17:240–250

Ertan KT, Hemleben V, Hemleben C (2004) Molecular evolution of some selected benthic foraminifera as inferred from sequences of the small subunit ribosomal DNA. Mar Micropaleontol 53:367–388

Fujita Y, Ito S (1957) A study of the foraminiferal assemblages from the Miocene formation, Date District, Fukushima Prefecture, Japan. J Geol Soc Jpn 63:497–513 (in Japanese, with English abstract)

Grell KG (1954) Der Generationswechsel der polythalamen Foraminifere *Rotaliella heterocaryotica*. Arch Protistk 100:211–235

Grell KG (1957) Untersuchungen über die Fortpflanzung und sexualität der Foraminiferen I. *Rotaliella roscoffensis*. Arch Protistk 102:147–164

Grell KG (1958) Untersuchungen über die Fortpflanzung und sexualität der Foraminiferen III. *Glabratella sulcata*. Arch Protistk 102:449–472

Grell KG (1979) Cytogenetic systems and evolution in foraminifera. J Foramin Res 9:1–13

Grimm GW, Stögerer K, Ertan KT, Kitazato H, Kučera M, Hemleben V, Hemleben C (2007) Diversity of rDNA in *Chilostomella*: molecular differentiation patterns and putative hermit types. Mar Micropaleontol 62:75–90

Guindon S, Dufayard JF, Lefort V, Anisimova M, Hordijk W, Gascuel O (2010) New algorithms and methods to estimate maximum-likelihood phylogenies: assessing the performance of PhyML 3.0. Syst Biol 59:307–321

Hallock PM, Cottey TL, Forward LB, Halas J (1986) Population biology and sediment production of *Archaias angulatus* (foraminiferida) in largo sound Florida. J Foramin Res 16:1–8

Hasegawa S (1979) Foraminifera of the Himi group, Hokuriku Province, Central Japan. The science reports of the Tohoku university, 2nd series (Geology), vol 49, pp 89–163

Higuchi Y (1954) Fossil foraminifera from the Miyata formation, Miura Peninsula, Kanagawa Prefecture. J Geol Soc Jpn 60:138–145 (in Japanese with English abstract)

Hirooka K (1988) Neogene paleoposition of the Japanese Islands inferred from paleomagnetic studies. In: Tsuchi R, Chiji M, Takayanagi Y (eds) IGCP project 246-Pacific Neogene events in time and space: Osaka museum of natural history, special publication. Osaka Museum of Natural History, Osaka, pp 3–16 (in Japanese, with English abstract)

Holzmann M (2000) Species concept in foraminifera: *Ammonia* as a case study. Micropaleontology 46(Suppl 1):21–37

Holzmann M, Pawlowski J (1996) Preservation of foraminifera for DNA extraction and PCR amplification. J Foramin Res 26:264–267

Holzmann M, Pawlowski J (2000) Taxonomic relationships in the genus *Ammonia* (Foraminifera) based on ribosomal DNA sequences. J Micropaleontol 19:85–95

Ishitani Y, Ishikawa SA, Inagaki Y, Tsuchiya M, Takahashi K, Takishita K (2011) Multigene phylogenetic analyses including diverse radiolarian species support the "Retaria" hypothesis: the sister relationship of Radiolaria and Foraminifera. Mar Micropaleontol 81:32–42

Ishitani Y, Kamikawa R, Yabuki A, Tsuchiya M, Inagaki Y, Takishita K (2012) Evolution of elongation factor-like (EFL) protein in rhizaria is revised by radiolarian EFL gene sequences. J Eukaryot Microbiol 59:367–373

Kaizuka S (1984) Landforms in and around the South Fossa Magna and their tectonic processes of growth. Quaternary Res 23:55–70 (in Japanese, with English abstract)

Kamikawa R, Inagaki Y, Sako Y (2008) Direct phylogenetic evidence for lateral transfer of elongation factor-like gene. Proc Natl Acad Sci USA 105:6965–6969

Kanesaki H (1987) Geographic distributions of rocky shore foraminifera adjacent to the Japanese Islands. Undergraduate thesis, Institute of Geosciences, Shizuoka University, P 86 (in Japanese with English abstract)

Kimura M (1980) A simple method for estimating evolutionary rates of base substitutions through comparative studies of nucleotide sequences. J Mol Evol 16:111–120

Kitazato H (1984) Microhabitats of benthic foraminifera and their application to fossil assemblages. In: Oertli HJ (ed) Benthos '83: 2nd international symposium on benthic foraminifera, France, pp 339–344

Kitazato H (1988) Ecology of benthic foraminifera in the tidal zone of a rocky shore. Benthos '86, pp 815–825

Kitazato H (1992) Pseudopodia of benthic foraminifera and their relationships to the test morphology. In: Takayanagi Y, Saito T (eds) Studies in benthic foraminifera. Tohoku University Press, Tokyo, pp 103–108

Kitazato H (1994) Foraminiferal microhabitats in four marine environments around Japan. Mar Micropaleontol 24:29–41

Kitazato H (1997) Paleogeographic changes in central Honshu, Japan, during the late Cenozoic in relation to the collision of the Izu-Ogasawara Arc with the Honshu Arc. Island Arc 6:144–157

Kitazato H, Tsuchiya M, Takahara K (2000) Recognition of breeding populations in foraminifera: an example using the genus *Glabratella*. Paleontol Res 4:1–15

Lee JJ, Faber WW Jr, Anderson OR, Pawlowski J (1991) Life–cycles of foraminifera. In: Lee JJ, Anderson OR (eds) Biology of foraminifera. Academic, London, pp 285–334

Matoba Y, Tomizawa A, Maruyama T, Shiraishi T, Aita Y, Okamoto K (1990) Neogene and quaternary sedimentary sequences in the Oga Peninsula. Benthos '90: the 4th international symposium on benthic foraminifera, Sendai, Japan

Matsushima Y (1984) Shallow marine molluscan assemblages of postglacial period in the Japanese islands: its historical and geographical changes induced by the environmental changes. Bull Kanagawa Pref Mus 15:37–109 (in Japanese, with English abstract)

Miya M, Nishida M (1997) Speciation in the open ocean. Nature 389(803):804

Moodley L, Hess C (1992) Tolerance of infaunal benthic foraminifera for low and high oxygen concentrations. Biol Bull 183:94–98

Murray JW (2006) Ecology and applications of benthic foraminifera. Cambridge University Press, Cambridge, 426 pp

Myers EH (1938) The present state of our knowledge concerning the life cycle of the foraminifera. Proc Natl Acad Sci USA 24:10–17

Myers EH (1940) Observations on the origin and fate of flagellated gametes in multiple tests of *Discorbis* (Foraminifera). J Mar Biol Assoc UK 24:201–226

Nakamura K, Shimazaki K, Yonekura N (1984) Subduction, bending and eduction: present and Quaternary tectonics of the northern border of the Philippine Sea plate. Bull Soc Géol France XXVI:221–243

Nemoto N, Yoshimoto N (2001) Foraminiferal fossils from the Pleistocene Hamada formation in the Chikagawa area, Eastern Shimokita Peninsula, Northeast Japan. Fossils (Palaeontol Soc Jpn) 69:1–24 (in Japanese, with English abstract)

Nomura R (1992) Miocene benthic foraminifers at sites 794, 795, and 797. Proc Ocean Drill Program Sci Results 127/128:493–540

Oba T, Kato M, Kitazato H, Koizumi I, Omura A, Sakai T, Takayama T (1991) Paleoenvironmental changes in the Japan Sea during the last 85,000 years. Paleoceanography 6:499–518

Ogasawara K (1994) Neogene paleogeography and marine climate of the Japanese Islands based on shallow marine molluscs. Palaeogeogr Palaeoclimatol Palaeoecol 108:335–351

Otofuji Y, Matsuda T (1984) Timing of rotational motion of southwest Japan inferred from paleomagnetism. Earth Planet Sci Lett 70:373–382

Pawlowski J (2000) Introduction to the molecular systematics of foraminifera. Micropaleontology 46(suppl 1):1–12

Pawlowski J, Holzmann M (2008) Diversity and geographic distribution of benthic foraminifera: a molecular perspective. Biodivers Conserv 17:317–328

Pawlowski J, Bolivar I, Fahrni JF, Cavalier-Smith T, Gouy M (1996) Early origin of Foraminifera suggested by SSU rRNA gene sequences. Mol Biol Evol 13:445–450

Pawlowski J, Fahrni J, Lecroq B, Longet D, Cornelius N, Excoffier L, Cedhagen T, Gooday AJ (2007) Bipolar gene flow in deep-sea benthic foraminifera. Mol Ecol 16:4089–4096

Pillet L, Fontaine D, Pawlowski J (2012) Intra-genomic ribosomal RNA polymorphism and morphological variation in *Elphidium macellum* suggests inter-specific hybridization in foraminifera. PLoS ONE 7:e32373. doi:10.1371/journal.pone.0032373

Rambaut A (1996) Se-Al version 2.0a11: sequence alignment editor. http://evolve.zoo.ox.ac.uk/

Ross CA (1972) Biology and ecology of *Marginopora vertebralis* (Foraminiferida), Great Barrier Reef. J Protozool 19:181–192

Röttger R (1974) Larger foraminifer: reproduction and early stages of development in *Heterostegina depressa*. Mar Biol 26:5–12

Röttger R, Krüge R, de Rijk S (1990) Trimorphism in foraminifera (protozoa)-verification of an old hypothesis. Euro J Protist 25:226–228

Saitou N, Nei M (1987) The neighbor-joining method: a new method for reconstructing phylogenetic trees. Mol Biol Evol 4:406–425

Sakaguchi M, Takishita K, Matsumoto T, Hashimoto T, Inagaki Y (2009) Tracing back EFL gene evolution in the cryptomonads-haptophytes assemblage: separate origins of EFL genes in haptophytes, photosynthetic cryptomonads, and goniomonads. Gene 441:126–131

Schweizer M, Pawlowski J, Duijnstee IAP, Kouwenhoven TJ, van der Zwaan GJ (2005) Molecular phylogeny of the foraminiferal genus *Uvigerina* based on ribosomal DNA sequences. Mar Micropaleontol 57:51–67

Spindler M (1980) The pelagic gulfweed *Sargassum natans* as a habitat for the benthic foraminifera *Planorbulina acervalis* and *Rosalina globularis*. Neues Jahrbuch Geol Paläont Mh 110:569–580

Takahara K (1989) Morphological analysis by plastogamy experiment on the benthic foraminifera, Glabratella. Undergraduate thesis, Institute of Geosciences, Shizuoka University, p 61 (in Japanese with English abstract)

Takishita K, Inagaki Y, Tsuchiya M, Sakaguchi M, Maruyama T (2005) A close relationship between Cerocozoa and foraminifera supported by phylogenetic analysis based on combined amino acid sequences of three cytoskeletal proteins (actin, α-tubulin, and β-tubulin). Gene 362:153–160

Tamura K, Peterson D, Peterson N, Stecher G, Nei M, Kumar S (2011) MEGA5: molecular evolutionary genetics analysis using maximum likelihood, evolutionary distance, and maximum parsimony methods. Mol Biol Evol 28:2731–2739

Thompson JD, Higgins DG, Gibson TJ (1994) CLUSTAL W: improving the sensitivity of progressive multiple sequence alignment through sequence weighting, position-specific gap penalties and weight matrix choice. Nucleic Acids Res 22:4673–4680

Toyofuku T, Kitazato H (2005) Micromapping of Mg/Ca values in cultured specimens of the high-magnesium benthic foraminifera. Geochem Geophys Geosyst 6:Q11P05. doi:10.1029/2005GC000961

Toyofuku T, Kitazato H, Kawahata H, Tsuchiya M, Nohara M (2000) Evaluation of Mg/Ca thermometry in foraminifera: comparison of experimental results and measurements in nature. Paleoceanography 15:456–464

Tsuchiya M, Aizawa M, Suzuki-Kanesaki H, Kitazato H (1994) Life history of *Glabratella opercularis* (d'Orbigny): observations and experiments. PaleoBios 16:62

Tsuchiya M, Kitazato H, Pawlowski J (2000) Phylogenetic relationships among species of Glabratellidae (Foraminifera) inferred from ribosomal DNA sequences: comparison with morphological and reproductive data. Micropaleontology 46(suppl 1):13–20

Tsuchiya M, Kitazato H, Pawlowski J (2003) Analysis of internal transcribed spacer of ribosomal DNA reveals cryptic speciation in *Planoglabratella opercularis*. J Foramin Res 33:285–293

Tsuchiya M, Tazume M, Kitazato H (2008) Molecular characterization of the non-costate morphotypes of buliminid foraminifers based on internal transcribed region of ribosomal DNA (ITS rDNA) sequence data. Mar Micropaleontol 69:212–224

Tsuchiya M, Grimm GW, Heinz P, Stögerer K, Ertan KT, Collen J, Brüchert V, Hemleben C, Hemleben V, Kitazato H (2009) Ribosomal DNA shows extremely low genetic divergence in a world-wide distributed, but disjunct and highly adapted marine protozoan (*Virgulinella fragilis*, Foraminiferida). Mar Micropaleontol 70:8–19

Ujiié Y, Asami T, de Garidel-Thoron T, Liu H, Ishitani Y, de Vargas C (2012) Longitudinal differentiation among pelagic populations in a planktic foraminifer. Ecol Evol. doi:10.1002/ece3.286

Chapter 10
Survival, Reproduction and Calcification of Three Benthic Foraminiferal Species in Response to Experimentally Induced Hypoxia

Emmanuelle Geslin, Christine Barras, Dewi Langlet, Maria Pia Nardelli, Jung-Hyun Kim, Jérôme Bonnin, Edouard Metzger, and Frans J. Jorissen

Abstract An experiment was conducted to test the survival rates, growth (calcification), and reproduction capacities of three benthic foraminiferal species (*Ammonia tepida*, *Melonis barleeanus* and *Bulimina marginata*) under strongly oxygen-depleted conditions alternating with short periods of anoxia. Protocols were determined to use accurate methods (1) to follow oxygen concentrations in the aquaria (continuously recorded using microsensors), (2) to distinguish live foraminifera (fluorogenic probe), (3) to determine foraminiferal growth (calcein-marked shells and automatic measurement of the shell size). Our results show a very high survival rate, and growth of *A. tepida* and *M. barleeanus* in all experimental conditions, suggesting that survival and growth are not negatively impacted by hypoxia. Unfortunately, no reproduction was observed for these species, so that we cannot draw firm conclusions on their ability to reproduce under hypoxic/anoxic conditions. The survival rates of *Bulimina marginata* are much lower than for the other two species. In the oxic treatments, the presence of juveniles is indicative of reproductive events, which can explain an important part of the mortality. The absence of juveniles in the hypoxic/anoxic treatments could indicate that these conditions inhibit reproduction. Alternatively, the perceived absence of juveniles could also be due to the fact that the juveniles resulting from reproduction (causing similar mortality rates as in the oxic treatments) were not able to calcify, and remained at a propagule stage. Additional experiments are needed to distinguish these two options.

E. Geslin (✉) • C. Barras • D. Langlet • M.P. Nardelli • E. Metzger • F.J. Jorissen
UNAM University, Université d'Angers, UMR CNRS 6112 LNPG-BIAF,
2 Bd Lavoisier, 49045 Angers Cedex, France
e-mail: emmanuelle.geslin@univ-angers.fr

J.-H. Kim
Department of Marine Organic Biogeochemistry, Royal Netherlands Institute for Sea Research (NIOZ), PO Box 59, 1790 AB Den Burg, Texel, The Netherlands

J. Bonnin
Université de Bordeaux 1, UMR 5805 EPOC, Avenue des Facultés,
33405 Talence Cedex, France

H. Kitazato and J.M. Bernhard (eds.), *Approaches to Study Living Foraminifera: Collection, Maintenance and Experimentation*, Environmental Science and Engineering,
DOI 10.1007/978-4-431-54388-6_10, © Springer Japan 2014

Keywords Anoxia • Benthic foraminifera • Growth • Hypoxia • Reproduction • Survival rate

10.1 Introduction

Oxygen deficiency is one of the most widespread harmful effects for aerobic organisms in the marine environment. Over the past 10 to 15 years the number of coastal areas affected by seasonal hypoxia in the bottom waters has spread rapidly, mainly due to anthropogenically induced eutrophication (Diaz and Rosenberg 1995; Diaz and Rosenberg 2008). While hypoxic and anoxic events (hypoxic<63 μM O_2, anoxic = below the detection limit of microsensors, following Middelburg and Levin 2009) existed throughout geological time, their occurrence in shallow marine and estuarine areas is clearly increasing. Global warming will probably enhance these effects and will enlarge the affected areas (Justic et al. 2003).

In most marine environments, only the top centimeter or millimeter of the sediment contain oxygen (Jørgensen and Revsbech 1989), because it is quickly consumed by biological and chemical processes (Wenzhöfer and Glud 2004). The oxygen content, which usually shows an exponential downward decrease, is the result of the equilibrium between downward oxygen diffusion from the bottom water into the sediment and the consumption of the oxygen used for the degradation of organic matter by aerobic organisms (Glud 2008). Oxygen deficiency may be caused by increased organic matter supplies and/or a decrease of the bottom water ventilation. At greater water depths, less organic matter reaches the seafloor, and less carbon is degraded in the sediment. Consequently, in general, in fined-grained organic-rich sediments in marine coastal areas, oxygen penetration is only a few millimeters (e.g. Jørgensen 2005) compared to several centimeters in the deep sea (Wenzhöfer and Glud 2004). In addition, other important factors controlling oxygen availability in sediments are photosynthetic benthic microorganisms (Revsbech et al. 1986), such as diatoms which form biofilms in the intertidal zone and burrowing macrofauna which increase oxygen penetration into the sediment (bio-irrigation) in deeper ecosystems (e.g. Aller 1994).

In environments with hypoxic bottom waters, oxic organic matter degradation is decreased but remineralization will continue using nitrates, sulfates and metal oxides. In permanently anoxic or sulfidic settings, most oxidants have been exhausted and early diagenetic processes in the sediments are dominated by sulfate reduction, methanogenesis and anaerobic oxidation of methane (Treude 2012).

Marine faunas are strongly affected by hypoxic and anoxic events, particularly the benthic fauna living at the sediment-water interface and within the superficial sediment. In general meiofauna is less sensitive to hypoxia and anoxia than macrofauna and megafauna (Josefson and Widbom 1988; Moodley et al. 1998). Obligate or facultative anaerobes occur in some meiofaunal protists groups including flagellates, ciliates and foraminifera (Fenchel 2012). The diversity of anaerobic metabolic pathways applied by eukaryotes is much more restricted than the wide range of processes that enable prokaryotes to live and grow in the absence of free oxygen. Still, new findings show from time to time that eukaryotic organisms have developed some unexpected strategies to exert all basic life functions (survival, growth, reproduction) under anaerobic conditions (Oren 2012). However, anaerobic respiration using oxidized compounds of nitrogen, sulfur and other elements remain very rare among eukaryotes. Following observations that foraminifera may proliferate in anoxic environments (reviews in Bernhard and Sen Gupta 1999; Geslin et al. 2004), Risgaard-Petersen et al. (2006) discovered that a foraminiferal species (*Globobulimina turgida*) that usually lives in the oxygen-free zone of the sediment can use nitrate as electron acceptor and perform full denitrification forming N_2 as end product. This species is able to accumulate a large quantity of nitrate in its cell, allowing it to employ anaerobic metabolism for at least 3 months (Piña-Ochoa et al. 2010b). Piña-Ochoa et al. (2010a) recently showed that many other benthic foraminiferal species store nitrate in their cell, suggesting that nitrate reduction can potentially be applied by a wide range of species. A gene for nitrate reduction has recently been found in *Bolivina argentea* (Bernhard et al. 2012), suggesting that in this species denitrification could at least partially be performed by the foraminifera themselves, and not entirely by symbiotic bacteria, such has been shown for allogromiid foraminifera (Bernhard et al. 2011). However, some foraminiferal species known to be tolerant to hypoxic and anoxic conditions do not store nitrate in their cell, strongly suggesting that they are not able to denitrify (Piña-Ochoa et al. 2010a). For example, this is the case of *Ammonia tepida*, a dominant shallow water species, often observed in deeper layers in hypoxic or anoxic sediments (e.g. Kitazato 1994; Bouchet et al. 2009). Also *Bulimina marginata* and *Melonis barleeanus,* which have often been observed in deeper sediment layers, and are therefore suspected to be facultative anaerobes, do not show elevated concentrations of nitrate in their cell (Piña-Ochoa et al. 2010a).

In the present paper, we want to focus on the response of these somewhat enigmatic species to alternating hypoxic and anoxic conditions. We were especially interested to determine their survival rate and growth capacity under hypoxic conditions and their ability to shift to nitrate reduction during short periods of anoxia.

To answer these questions, we studied in the laboratory the behavior of these three species (*Ammonia tepida*, *Bulimina marginata* and *Melonis barleeanus*) under different oxygen regimes, using adequate methods to distinguish live specimens, to observe the growth of living individuals, and to identify reproduction events. Three different treatments were applied: (1) oxic conditions with available nitrate, (2) strongly hypoxic to anoxic (nitrogen-flushed) conditions with nitrate and (3) strongly hypoxic to anoxic (nitrogen-flushed) conditions without nitrate. We expected that the third treatment would allow us to observe the ability of the three species to survive short periods of anoxia without having the possibility to store nitrate and/or to denitrify.

10.2 Ecological Characteristics of the Three Tested Species

10.2.1 Ammonia tepida

Ammonia tepida is a cosmopolitan species occurring in intertidal mudflats, brackish lagoons, estuaries and shallow marine areas that are extremely variable environments both temporally and spatially (Debenay et al. 2000; Murray 2006). It is a common species, which is able to survive a wide range of temperature, salinity and other environmental parameters varying on seasonal or daily scales (e.g. Bradshaw 1961; Schnitker 1974; Walton and Sloan 1990). It is a free living superficial to infaunal species in muddy to silty or fine sandy sediments (e.g. Debenay et al. 1998).

Ammonia tepida has been extensively studied for various attributes, such as geographic distribution, ecology, biology, life-cycles, morphology, structure, and dependence of environmental parameters such as temperature and salinity (e.g., Bradshaw 1957, 1961; Schnitker 1974; Poag 1978; Jorissen 1988; Walton and Sloan 1990; Goldstein and Moodley 1993; Geslin et al. 1998; Stouff et al. 1999a, b; Pascal et al. 2008). *Ammonia tepida* is considered as a deposit feeder and has been found feeding on algae (e.g. Moodley et al. 2000), bacteria (Goldstein and Corliss 1994; Langezaal et al. 2005; Pascal et al. 2008) and other meiofaunal groups (Dupuy et al. 2010). The congener *Ammonia* sp. shows a rapid uptake of freshly deposited algal carbon (Moodley et al. 2000).

The large morphological variability of the genus *Ammonia* has led to considerable difficulties in species identification and more than 40 species and subspecies (or varieties) of recent *Ammonia* have been described worldwide under the generic names *Ammonia, Streblus*, and *Rotalia* (Ellis and Messina 1940). Recent studies have shown that the total number of genetically distinct and morphologically separable living species of *Ammonia* worldwide is likely to exceed 25–30 (Hayward et al. 2004). Comparing the morphology of *Ammonia tepida* collected in the Aiguillon Bay with the pictures presented by Hayward et al. (2004), our morphological range contains morphotypes close to molecular type T1.

Ammonia tepida is reported to be able to live in oxic and hypoxic conditions (Moodley and Hess 1992; Koho and Piña-Ochoa 2012). It has been found both in epifaunal and infaunal microhabitats (Kitazato 1994; Debenay et al. 1998). Bouchet et al. (2009) found Rose Bengal stained specimens below 20 cm in the sediment. However, it is still not 100 % clear whether this species is able to live in anoxic sediments, and, if this is the case, whether it uses an anaerobic metabolism or if it is able to survive in oxic microniches created by bioturbation. We expected our experiment to confirm or discount the ability of *A. tepida* to survive and to calcify under strongly hypoxic conditions lasting for two months and to clarify its capacity to denitrify during short periods of anoxia.

10.2.2 *Bulimina marginata*

Bulimina marginata is a cosmopolitan species found in various oceans around the world (e.g. Jorissen 1988; Corliss 1991; Jannink et al. 1998; Jorissen et al. 1998; De Rijk et al. 1999, 2000; Schmiedl et al. 2000; Mojtahid et al. 2006, 2008, 2010; Mojtahid 2007; Pucci et al. 2009). *Bulimina marginata* lives at a wide range of water depths, from the continental shelf between 10 and 150 m water depth (e.g. Barmawidjaja et al. 1992; Fontanier et al. 2002; Langezaal et al. 2005) to continental slope and bathyal environments, down to 2,200 m water depth (e.g. Jorissen et al. 1998; De Rijk et al. 1999, 2000). Abundant faunas of this species have also been described in submarine canyon environments (Schmiedl et al. 2000; Hess et al. 2005; Koho et al. 2007; Hess and Jorissen 2009).

Several studies indicate that *B. marginata* is a eutrophic taxon typical of fine grained muddy sediments with high food availability (e.g., Jorissen 1988; De Rijk et al. 2000). *Bulimina marginata* is generally considered as a detritivore. It occurs alive in shallow infaunal microhabitats in the topmost centimeters of the sediment, but abundant Rose Bengal stained assemblages have also been described in much deeper sediment layers (e.g. Kitazato 1989; Jorissen et al. 1998; Jorissen 1999). The presence of such deep assemblages may be related to macrofaunal burrows.

In several older studies, *B. marginata* has been considered as a good marker of low oxygen conditions (e.g. Phleger and Soutar 1973; van der Zwaan and Jorissen 1991; Sen Gupta and Machain-Castillo 1993; Bernhard and Sen Gupta 1999). It is supposed to support periodic episodes of anoxia or hypoxia (Pucci et al. 2009).

10.2.3 *Melonis barleeanus*

Melonis barleeanus is a typical species of open marine settings from the lower continental shelf to bathyal environments (e.g. Schmiedl et al. 2000; Fontanier et al. 2002, 2008). Rich assemblages have also been described in submarine canyons (Koho et al. 2008; Nardelli et al. 2010; Phipps et al. 2012).

Melonis barleeanus is commonly described as an intermediate infaunal species (Jorissen et al. 1998). This species is most often found in hypoxic sediments where pore-water nitrate concentrations are maximal (Jorissen et al. 1998; Fontanier et al. 2005; Koho et al. 2008; Mojtahid et al. 2010), and is assumed to feed on degraded organic detritus (e.g. Caralp 1989; Fontanier et al. 2002). Although only very low nitrate concentrations have been measured in *Melonis barleeanus*, (intracellular NO_3^- content: 0.6 ± 0.2 mM; Piña-Ochoa et al. 2010a), it has been considered as a potential denitrifier because of its systematic appearance in nitrate-rich sediment layers (Koho and Piña-Ochoa 2012).

10.3 Methodology

10.3.1 Experimental Design

10.3.1.1 Foraminiferal Collection and Storage in the Laboratory

Ammonia tepida

Superficial sediment was sampled at low tide on the 9th of February 2011 in the intertidal area of the Bay of Aiguillon (French Atlantic coast). This area is known for its very rich live faunas of *Ammonia tepida* (Pascal et al. 2008). In the field, the sediment was immediately sieved, using a 150-μm mesh and local seawater. The fraction >150 μm, containing large amounts of foraminifera, was transported to the culture laboratory at the University of Angers and maintained at 14 °C until the beginning of the experiment. On the 15th of April 2011 (one month before the beginning of the experiment), *Ammonia tepida* was incubated in a calcein solution (Bernhard et al. 2004).

Bulimina marginata

Sediment cores were sampled in the Bay of Biscay at station K (Lat. 43°38′N, Long. 1°43′W, water depth 650 m depth) on the 4th of February 2011 with a classical Barnett multi-tube corer (Barnett et al. 1984). The two first centimeters were sliced from the subsurface, placed in containers with seawater and transported to the laboratory where they were stored at 10 °C until the beginning of the experiment.

Melonis barleeanus

Sediment cores were sampled with a MUC8+4 multi-tube corer in Nazaré Canyon at station 161 (Lat. 39°35′N, Long. 9°24′W, water depth 918 m) on the 20th of March 2011. The two first centimeters were sliced from the subsurface, placed in containers with seawater and transported to the laboratory where they were stored at 12 °C until the beginning of the experiment.

10.3.1.2 Identification and Collection of Living Organisms

One week before the start of the experiment, sediments from the 3 different locations were sieved with filtered (0.45 μm) seawater over a 150-μm mesh and living foraminifera were picked. To ascertain the vitality of the selected specimens, each foraminifer was placed in a thin layer of fine-grained (<38 μm) sediment for approximately 12 h. Only the specimens that had moved, and had produced a burrow or trace on the sediment were considered alive and selected for the experiment.

10.3.1.3 Experimental Set-Up

Three different conditions were tested in three different aquaria with the same temperature (12.0 ± 0.3 °C) and salinity (35.4 ± 0.2).

Aquarium "OxN" (Oxic + Nitrate) contained well oxygenated artificial seawater (ASW) with 50 μM NO_3. High oxygen concentrations were maintained by continuously bubbling seawater with air. Aquarium "HypN" (low [O_2] + Nitrate) and aquarium "Hyp" (low [O_2] without nitrate) were both maintained under low oxygen conditions and contained ASW with 50 μM NO_3 for HypN or nitrate-free water for Hyp. Hypoxic conditions were obtained by a continuous introduction of N_2 gas mixed with 0.04 % CO_2 (to keep the pH stable (Piña-Ochoa et al. 2010b)) into the overlying water. Oxygen concentrations were measured at least ones a day with a Clark-type microelectrode with a 100 μm thick tip (Revsbech 1989). The sensors were regularly 2-point calibrated in air-saturated ASW (temperature 12 °C, salinity 35) and in anoxic alkaline ascorbate.

Specimens were introduced into the aquaria in glass vials (20 mL each) with a thin layer of fine-grained sieved sediment (<38 μm) filled up with ASW, in order to create more 'natural' conditions for the foraminifera. Each vial was closed with a net (100 μm mesh) to keep the foraminifera within the vials.

10.3.1.4 Survival Experiment

Living adult specimens of the three selected species *Ammonia tepida, Melonis barleeanus* and *Bulimina marginata* were introduced in 45 species-specific glass vials (one vial per sampling time per species per aquarium): 20 specimens per vial for *A. tepida*, 20 for *M. barleeanus* and 11 for *B. marginata*. Each aquarium contained 5 vials with *A. tepida*, 5 vials with *M. barleeanus* and 5 vials with *B. marginata*. The vial position in the aquaria was randomly defined to avoid any spatial effect on the foraminiferal survival rates.

The experiment started on the 16th of May 2011 and ran for 56 days. Three vials (one per species) were sampled in each aquarium after 7 days (T = 1), 14 days (T = 2), 29 days (T = 3), 45 days (T = 4) and 56 days (T = 5). The contents (sediment + foraminifera) of the vials were transferred to Petri dishes and foraminifera were picked out. Nets and vials were checked under a stereomicroscope to pick also the specimens that had migrated upward (along the walls of the vials).

The majority of introduced foraminifera were recovered at the end of the experiment (80–100 % with one exception of 61 % at T = 3 in the HypN condition for *B. marginata*). In some cases, juveniles were born during the experiment; they have been counted as well. Living specimens were recognized using a fluorescence technique. The fluorescence technique is a powerful, reliable and quick method to identify living individuals in experimental setups (Bernhard et al. 1995; Pucci et al. 2009; Morigi and Geslin 2009; Piña-Ochoa et al. 2010b; Koho et al. 2011; Heinz and Geslin 2012), particularly under anoxic conditions. Incubation in fluorogenic probes causes live specimens to become fluorescent. The fluorogenic probe used in this experiment was fluorescein diacetate (FDA; Bernhard et al. 1995), which was diluted in a dimethyl sulfoxide solution and ASW in order to obtain a final solution of 100 μM FDA. Specimens were picked out at the end of experiment directly in the vials when it was possible or after sieving (>63 μm). They were incubated in ASW-FDA solution for 10–20 h. Foraminifera were then picked out and placed in ASW with no FDA. Specimens were inspected individually for fluorescence using an Olympus SZX12 with a light fluorescent source Olympus URFL-T. Survival rates were calculated for each species and for each experimental condition.

10.3.1.5 Growth Experiment

To study their growth ability under the various experimental conditions, one vial per species was added to each aquarium (Ox, HypN, Hyp). For *A. tepida* and *B. marginata*, calcein-marked living specimens were added; 50 semi-adults of *A. tepida* at T=0 and 20 juveniles (3 chambers) of *B. marginata* at T=1. Thanks to the calcein stain of the introduced tests, the newly formed chambers, calcified under controlled conditions, could be easily identified (Bernhard et al. 2004; Barras et al. 2009) and counted. Additionally, at the end of the experiment, all specimens were measured using an automatic particle analyzer (see description below). Specimens that did not calcify new chambers were used to estimate the initial size whereas specimens that calcified new chambers allowed us to determine the average size increase of *A. tepida* and *B. marginata* in each condition.

For *M. barleeanus*, specimens were not calcein-marked beforehand. To evaluate their growth during the experiments, the size of all introduced specimens (10 adult individuals per condition) was measured at the start (T=0) and the end (T=5) of the experiment using an automatic particle analyzer (see below).

10.3.2 Size Measurements Using an Automatic Particle Analyser

Automated particle analysis has been carried out with a fully automated incident light microscope system. Images are acquired and particles are analyzed with analySIS FIVE (SIS/Olympus©) software supported by a software add-in developed by MAS©. Samples are measured automatically under a Leica Z16APO monocular microscope (plan-apochromatic objective). Images are captured with a CC12 colour camera (SIS©). Constant illumination of samples is provided by a Leica© CLS100X light source and a Leica© ring-light. This setup is the second generation of a particle analyzing system developed by Jörg Bollmann (Bollmann et al. 2004; Movellan et al. 2012).

10.3.3 Metabolism: Oxygen Respiration Rates

In order to estimate the amount of oxygen needed for foraminiferal aerobic metabolism, respiration rates of *M. barleeanus* were measured using a Clark type O_2 microsensor (Revsbech 1989) according to the methodology of Geslin et al. (2011), so that our results are totally comparable to their data. Four measurements were performed for *M. barleeanus* with 4 to 7 specimens per analysis. Aerobic respiration rates of *A. tepida* from Aiguillon Bay and *B. marginata* from Bay of Biscay were not measured because data are available in Geslin et al. (2011).

10.3.4 Internal Nitrate Contents in Foraminifera

In order to follow the intracellular nitrate contents in the foraminiferal cells during the experiment, living specimens of the 3 studied species were selected in each condition at each sampling time (except for *A. tepida* at T=4 and T=5). One specimen was required for intracellular nitrate measurement. A total of 147 specimens were analyzed with 10–24 specimens per condition and per time. Each specimen was rinsed in nitrate-free ASW, transferred to PCR tubes and stored at −20 °C. They were analyzed for nitrate content using the VCl_3 reduction method (Braman and Hendrix 1989) on a chemiluminescence detector (Model CLD 86, Eco Physics AG) as described by Risgaard-Petersen et al. (2006) and Høgslund et al. (2008). Biovolumes were determined according to Geslin et al. (2011) in order to calculate the nitrate concentration in each individual.

10.3.5 Statistical Analyses

Two different parametric statistical procedures were used to test the effect of the experimental conditions (OxN, HypN and Hyp) and of time on the foraminiferal survival rates and growth. Due to the binomial nature of the response variable "survival" (i.e. dead or alive; 0 or 1) we used generalized linear model (GLM, Nelder and Wedderburn, 1972) procedures of R (R Development Core Team, 2011). In these GLM procedures we tested the effect of the quantitative independent variable "Time" and the qualitative independent variables "Conditions" (OxN, HypN and Hyp) and "Species" (*M. barleeanus*, *A. tepida* and *B. marginata*). This procedure allowed us to determine for each of the variables whether it had a significant effect on survival (expressed as a Deviance and a Chi probability) and if so, to quantify the effect (giving an estimate of a coefficient for which the significance was expressed with z and p statistics).

Because of the normality of the response variable (the dependant variable), we used linear models (LM, Chambers 1992) to quantify the effect of both "Conditions" and "Time" and their interaction on the size of the foraminifera (the Maximum Diameter or Length in μm according the species). In order to identify the effect of the conditions on each species we ran three models, one for each species. This procedure consists of two steps. The first one allows us to determine whether one or several independent variables (depending on the species: Time, Conditions or the Time*Conditions interaction) has a significant effect on the individual size (after an analysis of variance, expressed with F and the associated probability statistics). The aim of the second step was to quantify how each variable affects the foraminiferal size (expressed with an estimated coefficient, for which a Student's t-test is realized to see if the estimate is significantly different from zero, associated with a t and its probability statistics). As the method used to determine the growth is slightly different for the three species we have to use different independent variables for each species. Both for *A. tepida* and *B. marginata,* the initial size was not measured at the beginning of the experiment but was estimated on the basis of the final size of the individuals that did not grow during the experiment. For these two species, we tested the effect of the qualitative dependent variable coded with 4 terms: "T5.NoGrowth", "T5.OxN", "T5.HypN" and "T5.Hyp". For *M. barleeanus*, we tested the effect of Time, Conditions and of the interaction Time*Conditions on the dependant variable Maximum Diameter. For all procedures we used a 0.05 significance level.

10.4 Results

10.4.1 Monitoring of the Oxygen Conditions

During all the experiment in aquaria Hyp and HypN, oxygen concentrations were lower than 90 μmol L^{-1} (~2 mL L^{-1}) and several short anoxic periods occurred (up to 6 days). Average O_2 concentrations are 10 ± 13 and 14 ± 18 μmol L^{-1} in aquaria HypN and Hyp, respectively.

10.4.2 Survival Rates of Adult Specimens

10.4.2.1 Influence of Incubation Time Under Oxic Conditions

Under oxic conditions (OxN), the survival rates of *A. tepida* (84–100 %) and *M. barleeanus* (87–95 %) are very high; whereas the survival rate of *B. marginata* (36–64 %) is much lower (Figs. 10.1, 10.2 and 10.3). GLModels show that there is a significant "Species" effect with *B. marginata* exhibiting a lower survival rate (Table 10.1). There is no significant effect of the incubation time and of the interaction Time × Species on the survival rates of the three species (Table 10.1).

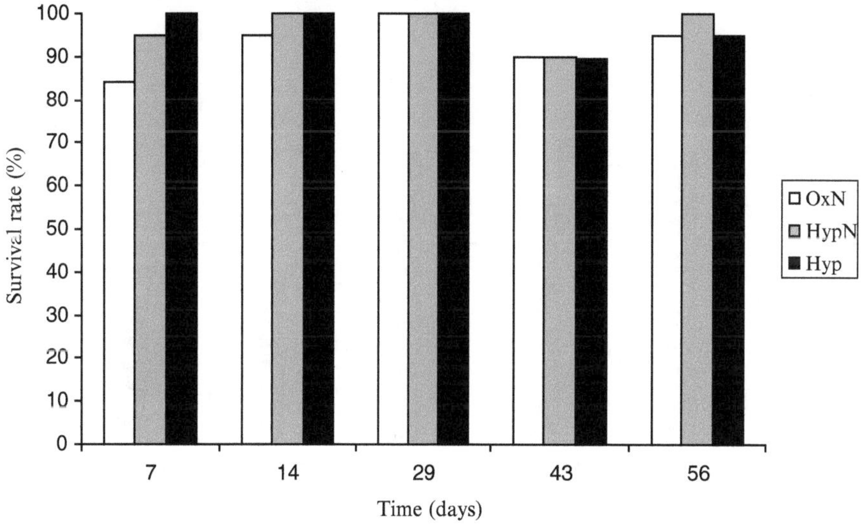

Fig. 10.1 Survival rates of *Ammonia tepida* after 7, 14, 29, 43 and 56 days under three experimental conditions

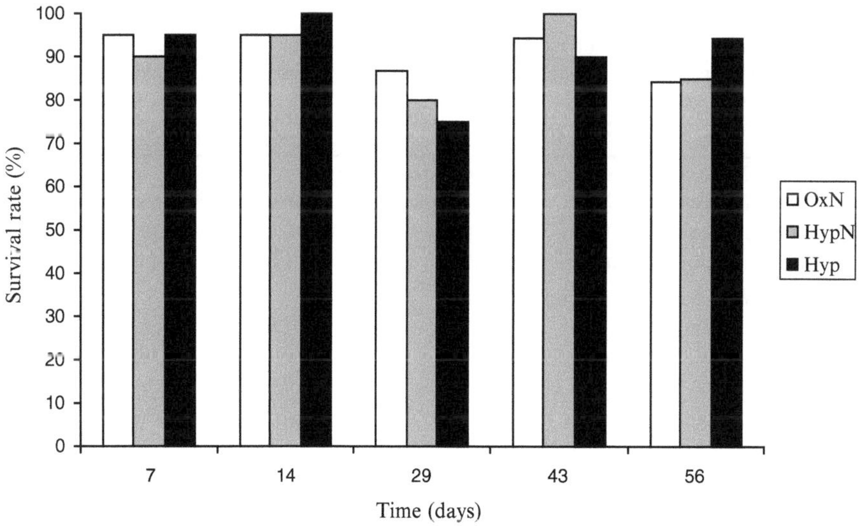

Fig. 10.2 Survival rates of *Melonis barleeanus* after 7, 14, 29, 43 and 56 days under three experimental conditions

10.4.2.2 Influence of Hypoxic Conditions

Specimens of the three tested species were found alive in both hypoxic conditions, until the end of the experiment. The survival rates (SR) are very high in both tested conditions for *A. tepida* (Fig. 10.1, 90–100 % for HypN, 89–100 % for Hyp,

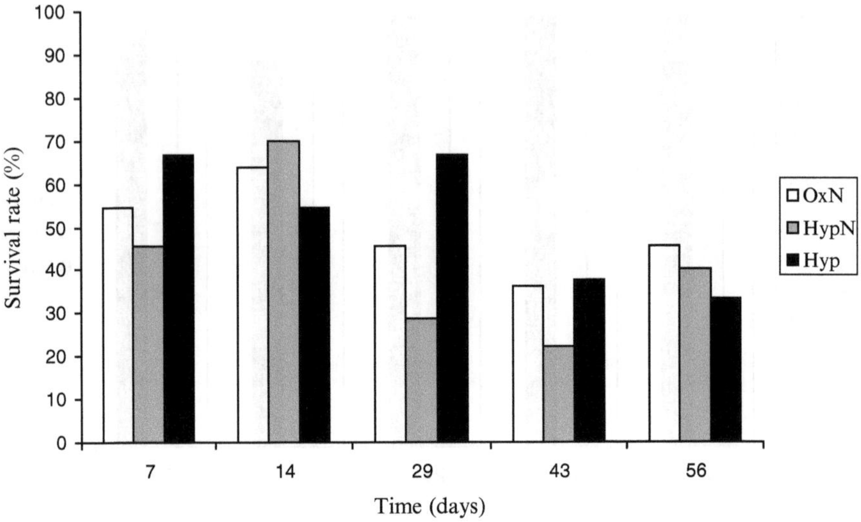

Fig. 10.3 Survival rates of *Bulimina marginata* after 7, 14, 29, 43 and 56 days under three experimental conditions

Table 10.1 Effect of the experimental conditions on survival rates of the three species under oxic conditions (upper panel) and under all conditions (lower panel)

	Df	Deviance	p(Chi)
Influence of incubation time under oxic conditions			
Species effect	2	46.9	6.6×10^{-11}
Incubation time effect	1	0.7	0.39
Interaction Incubation time × Species	2	1.8	0.40
Influence of hypoxic conditions			
Species effect	2	151.9	$<2 \times 10^{-16}$
Incubation time effect	1	2.7	0.10
Conditon effect	2	0.2	0.92

The Degrees of freedom (Df), the Deviance and the probability (p(Chi)) statistic values are presented for every tested variables

respectively) as well as for *M. barleeanus* (Fig. 10.2, 80–100 % for HypN, 75–100 % for Hyp) (Table 10.1). For *B. marginata*, the survival rates are significantly lower than for the other two species (22–70 % for HypN, 33–67 % for Hyp, respectively) (Table 10.1, Fig. 10.3). None of the three studied species exhibit significant differences in their survival rates between the different sampling times in hypoxic conditions alternating with short (up to 6 days) periods of anoxia with or without nitrates. No significant difference is observed between the aquaria with and without nitrate (Table 10.1).

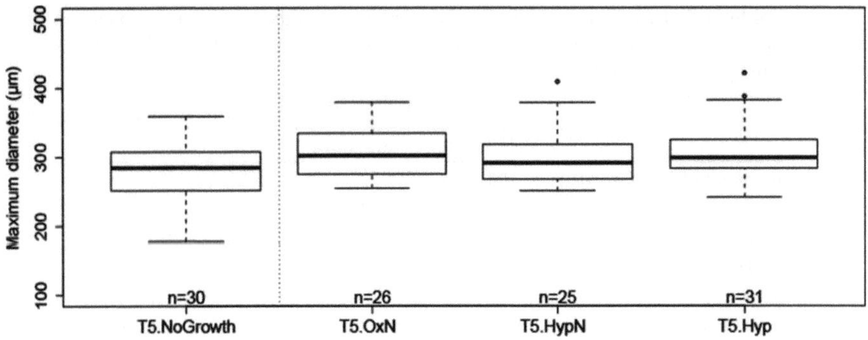

Fig. 10.4 Growth of *Ammonia tepida*: comparison of the size of the individuals that did not add new chambers (T5.NoGrowth) and those that calcified one or more new chambers (T5), for each tested condition (OxN, HypN, Hyp). On the box-and-whisker plot the two borders of each box are the first and third quartile while whiskers are 1.5 times the interquartile range of the lower and higher quartiles

10.4.3 Reproduction

No juveniles were obtained for *A. tepida* and *M. barleeanus,* whereas reproduction events occurred in the vials containing adult specimens of *B. marginata*. Juveniles were only found in the OxN aquaria: 80, 37 and 142 juveniles were counted at T=1, T=3 and T=4 respectively.

10.4.4 Growth

At T=5 (56 days), 64–84 % of the *A. tepida* semi-adult specimens introduced in the vials for the growth experiment were found. For this species, growth was assessed through the observation of additional chambers calcified during incubation (use of calcein labeling). The percentage of specimens that added at least one chamber in OxN, HypN and Hyp aquaria was 81 %, 60 % and 82 %, respectively (n=50 for each condition). They added 1 or 2 chambers with the exception of a single specimen that added 3 new chambers. In order to study the size increase for *A. tepida*, the average size of all the individuals that did not add any new chambers was compared with the average size of individuals that calcified new chambers (Fig. 10.4).

According to the size measurements, the individuals of *A. tepida* that added new chambers were significantly wider than the ones that did not exhibit newly formed chambers (Df=3, F=3.05, p=0.03). The average difference of maximum diameter between specimens that added new chambers and those who did not is +27.4, +17.5 and +28.2 µm in OxN, HypN and Hyp conditions, respectively (Fig. 10.4, Table 10.2). The size difference between individuals with and without chamber addition was significant in both the OxN and Hyp aquaria, but was not significant in the HypN aquaria (Table 10.2).

Table 10.2 Effect of the experimental conditions on the maximum diameter of *A. tepida*: estimation of the coefficient, its standard error (Std. error) and the t and p statistics for the linear model

	Maximum diameter (µm)		Statistical parameter	
	Estimate	Std. error	t	p
Intercept (T5.NoGrowth)	280.0	7.5	37.6	$<2.1\times10^{-16}$
T5.OxN	27.4	10.9	2.5	0.01
T5.HypN	17.5	11.1	1.6	0.12
T5.Hyp	28.2	10.5	2.7	0.01

Note that the estimated Intercept corresponds to average of the reference category (here T5.NoGrowth) whereas all other estimated values are indicative of the average difference between the experimental and the reference population

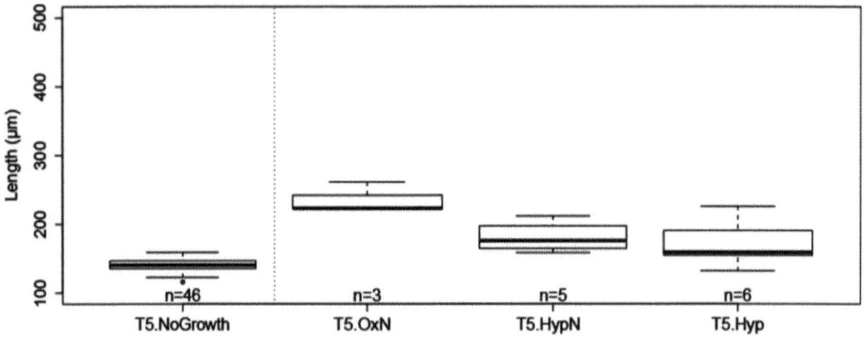

Fig. 10.5 Growth of *Bulimina marginata*: comparison of the size of the individuals that did not add new chambers (T5.NoGrowth) and those that calcified one or more new chambers (T5), for each tested condition (OxN, HypN, Hyp). On the box-and-whisker plot the two borders of each box are the first and third quartile while whiskers are 1.5 times the interquartile range of the lower and higher quartiles

All specimens of *B. marginata* introduced in the aquaria at T=1 (7 days) for the growth experiment (20 three-chambered juveniles for each treatment) were found at T=5 (56 days). The percentage of specimens that added at least one chamber in OxN, HypN and Hyp aquaria was 15, 25 and 30 %, respectively. Although only three specimens calcified in oxic conditions, each of these specimens added 2 or 3 new chambers. Conversely, the 11 specimens from hypoxic aquaria (5 in HypN and 6 in Hyp) that added new chambers added a single chamber, with the exception of a single specimen that added 2 chambers. In all three experimental conditions, at T=5, specimens that added new chambers had a significantly larger maximum length than specimens that did not add new chambers (Fig. 10.5, Df=3, F=49.7, p=8.35e-16 and Table 10.3). The average difference in maximum length between specimens with and without chamber addition was 95.7, 41.7 and 30.0 µm for OxN, HypN and Hyp, respectively (Table 10.3). A complementary statistical analysis shows that the size of the individuals that grew in the OxN conditions is significantly

Table 10.3 Effect of the experimental conditions on the maximum diameter of *B. marginata*: estimation of the coefficient, its standard error (Std. error) and the t and p statistics for the linear model

	Maximum diameter (μm)		Statistical parameter	
	Estimate	Std. error	t	p
Intercept (T5.NoGrowth)	140.0	2.2	62.9	$<2.1\times10^{-16}$
T5.OxN	95.7	9.0	10.6	4.5×10^{-15}
T5.HypN	41.7	7.1	5.9	2.5×10^{-7}
T5.Hyp	30.0	6.6	4.6	2.6×10^{-5}

Note that the estimated Intercept corresponds to average of the reference category (here T5.NoGrowth) whereas all other estimated values are indicative of the average difference between the experimental and the reference population

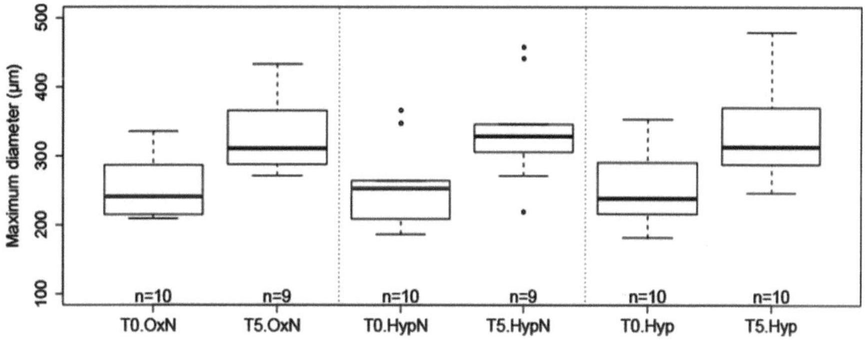

Fig. 10.6 Growth of *Melonis barleenus* between T0 and T5 for each tested condition (OxN, HypN, Hyp). On the box-and-whisker plot the two borders of each box are the first and third quartile while whiskers are 1.5 times the interquartile range of the lower and higher quartiles

higher than the size of the individuals that grew in both hypoxic conditions. Summarizing, it appears that hypoxic conditions do not have a negative effect on the number of specimens that add new chambers. Conversely, for specimens that added new chambers, more chambers were added in oxic conditions, and consequently, the size difference was larger. Finally, the presence/absence of nitrate in the seawater has no significant impact on chamber addition.

All 10 incubated specimens of *M. barleeanus* were found in the Hyp aquarium, whereas 9 from the 10 individuals were found in the OxN and HypN aquaria. In the case of *M. barleeanus*, the same specimens were measured at T=0 and T=5, so that the obtained values can be directly compared. In all three experimental conditions, the size increase between T=0 and T=5 (Fig. 10.6) was statistically significant (Df=1, F=25.2, p=6.32.10^{-6}), with an average increase of 81.6 μm, 85.9 μm and 80.3 μm for OxN, HypN and Hyp, respectively. No significant difference was found between the three experimental conditions (OxN, HypN, Hyp) (Df=2, F=0.05, p=0.95). It appears that, just as for *A. tepida*, strong hypoxic conditions alternating with short periods of anoxia do not inhibit the growth of *M. barleeanus*.

Table 10.4 Respiration rates (average value and standard error of the replicate measurements) of benthic foraminifera studied in the present paper with detailed data on their size (average size of maximal diameter or length and estimated biovolume)

Species	Location	Average size of max. Diam/Length	Biovolume (μm^3): Mean (SE)	RR (nL O2 $cell^{-1}$ h^{-1}) Mean	SE	n	References
Melonis barleeanus	Portugal Continental Margin	347	8.6E+06 (7.2E+06)	0.32	0.26	3	Our data
Ammonia tepida	Aiguillon Bay	541	3.1E+07 (3.2E+06)	2.01	0.07	4	Geslin et al. (2011)
Bulimina marginata	Bay of Biscay	369	8.0E+06 (6.4E+05)	0.42	0.07	3	Geslin et al. (2011)

10.4.5 Respiration Rates

Oxygen respiration measurements were performed for *M. barleeanus*. The observed values were 0.32 nL ± 0.26 O_2 $cell^{-1}$ h^{-1} (Table 10.4).

10.4.6 Intracellular Nitrate Content in Foraminifera

In the following part, data of intracellular nitrate content of the 3 studied species are introduced without taking into account the different tested conditions. In a second part, data are presented as a function of time for each species.

Nitrate contents were measured in 29, 57 and 61 specimens of *A. tepida*, *M. barleeanus* and *B. marginata,* respectively. Nitrate was detected within the cell of individuals of all three species, but about 40 % of the measured *B. marginata* and *A. tepida* individuals show no measurable nitrate in their cell (Table 10.5). The average values of nitrate contents are 16 ± 5, 61 ± 10 and 40 ± 7 pmol NO_3^- per cell for *A. tepida, M. barleeanus* and *B. marginata* respectively. The values of intracellular nitrate contents show a large intra-specific variability. For example, the average value of all measured *A. tepida* is 16 pmol NO_3^- $cell^{-1}$ with a standard error of the mean (SEM) of 5 pmol $cell^{-1}$, a minimum value of 0 and a maximum value of 114 pmol $cell^{-1}$ (n = 31) (Table 10.5). Taking into account the biovolume of each specimen, the concentrations in the cell can be calculated. The average concentrations of intracellular nitrate are varying between 3 and 12 mM according to the species (Table 10.5). These values are much higher than in the surrounding environments (max. 50 μM NO_3^-), underlining the ability for each of these three species to store nitrate.

Figure 10.7 shows the evolution of nitrate contents with time for each condition and each species. No significant trend is detectable except for *B. marginata* in the hypoxic aquarium without additional nitrate ($p < 0.05$). In this case, nitrate contents decreased with time.

Table 10.5 Intracellular nitrate contents recorded in the present study compared to the previous published data for the same species. n is the number of measured specimens

			Nitrate Content (pmol per cell)				
Species	Reference	Location	Min	Max	Average	NO_3^- (mM)[a]	n
Ammonia tepida	Our data	Aiguillon Bay	0	114	16 (5)	3 (1)	29
Melonis barleeanus	Our data	Portugal Margin	0	462	61 (10)	8 (1)	57
Bulimina marginata	Our data	Bay of Biscay	0	220	40 (7)	12 (2)	61
Ammonia tepida	Piña-Ochoa et al. (2010a, b)	Aiguillon Bay	0	58	13	0	15
Melonis barleeanus	Piña-Ochoa et al. (2010a, b)	North sea	1	27	9	1	7
Melonis barleeanus	Piña-Ochoa et al. (2010a, b)	Rhone delta	0	0	0	0	2
Bulimina marginata	Piña-Ochoa et al. (2010a, b)	Bay of Biscay	40	60	40	4	14

[a]Values are mean values. Standard Error of the Mean (SEM) is given in parentheses

10.5 Discussion

10.5.1 Species Responses to Hypoxic Conditions

The high survival rates obtained under oxic conditions during long-term laboratory incubation for *Ammonia tepida* are not really surprising. In fact, this species is known to be well adapted to laboratory conditions (e.g. Bradshaw 1957, 1961; Schnitker 1974; Goldstein and Moodley 1993; Schmidt et al. 1957; Stouff et al. 1999a, b; Le Cadre et al. 2003; Dissard et al. 2010a, b). According to the literature, this species is resistant to various types of environmental stress (Murray 2006). *Ammonia tepida* is a common species in intertidal mud-flats where the oxygen penetration depth is very low (few millimeters). Rose Bengal stained specimens have been observed either on top of the sediment (e.g. Alve and Murray 2001) or deeper in the sediment where conditions are mainly anoxic (Frankel 1975; Buzas 1977; Kitazato 1994; Bouchet et al. 2007). Consequently, *A. tepida* is often considered as a species resistant to oxygen-depleted conditions (Moodley and Hess 1992). However, it is still not clear if *A. tepida* inhabits the anoxic part of the sediment or occupies oxic microhabitats created by burrowing macrofauna. Our experiment confirms the capacity of *A. tepida* to survive hypoxic conditions (at least 2 months) alternating with short anoxic periods (up to 6 days).

Similar experimental observations were reported by Moodley and Hess (1992) for *Ammonia beccarii*. They observed the activity of this species after 6 days of incubation under low oxygen conditions (12 μM) and 24 h of anoxia. After a longer anoxic incubation experiment, Moodley et al. (1997) noted that specimens of the genus *Ammonia* were found alive (Rose Bengal stained) after 78 days of putative anoxia.

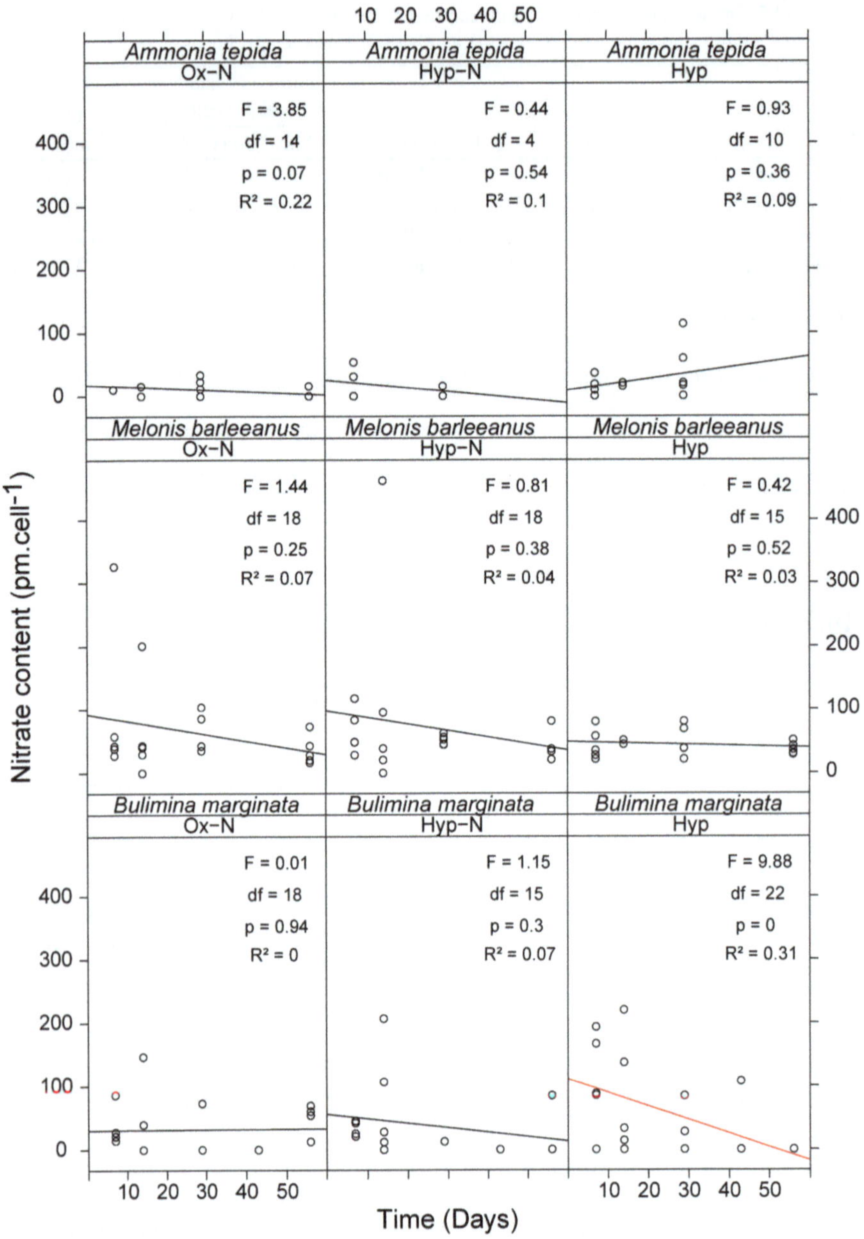

Fig. 10.7 Nitrate content (pmol cell^{-1}) recorded in the cell of the three species along the incubation time. The *red line* is statistically significant (only the case for *B. marginata*, Hyp)

However, using the Rose Bengal staining method in experimental studies is problematic; the number of living individuals may be totally unrealistic, since cytoplasm may stain for prolonged periods of time after the death of the organism, especially in anoxic conditions (Bernhard 1988).

Ammonia tepida is also known to be able to add chambers under laboratory condition (e.g. Bradshaw 1957, 1961; Goldstein and Moodley 1993; Stouff et al. 1999a, b; de Nooijer et al. 2009). In our case, growth was proven in all tested conditions by the observations of additional chambers not marked by calcein. Bradshaw (1961) reported growth rates of *Ammonia tepida* collected mainly in mud-flats (California, USA). He noted a growth of approximately 40 μm for adult specimens after 60 days at 15 °C. The values presented by Bradshaw (1961) are of the same order of magnitude as ours. Under oxic conditions, specimens of *A. tepida* that added new chambers (lack of calcein marking) were, on average, 27 μm larger in diameter than specimens that did not add chambers after 56 days of incubation. Our slightly lower values may be due to the fact that unlike Bradshaw (1961), we did not add food before or during the experiment. However, according to the large range of morphological and molecular types of *Ammonia tepida*, the specimens studied by Bradshaw (1961) were not necessarily conspecifics with our specimens.

Asexual reproduction has often been observed in *A. tepida* under laboratory conditions (Bradshaw 1957, 1961; Schnitker 1974; Goldstein and Moodley 1993; Stouff et al. 1999a; Diz et al. 2012). However, in our case, no reproduction (production of juveniles) was observed, not even in the oxic treatment. This may be due to low food availability in our experimental setup. In fact, Bradshaw (1961) has experimentally shown that *A. tepida* reproduces when food is added. Because of this absence of reproduction under oxic conditions, the results are inconclusive.

According to the literature, this is the first time that *M. barleeanus* has been used for an experimental study. The high survival rates obtained under oxic conditions show the ability of this species to survive under laboratory conditions. This result is very promising for future laboratory experiments using *M. barleeanus*. The majority of the specimens also survive 56 days of strong hypoxia alternating with short anoxic periods, with or without added nitrates. *Melonis barleeanus* is one of the most typical intermediate infaunal foraminiferal species (Jorissen 1999). It normally lives in a specific microhabitat at some millimeters or centimeters in the sediment, where oxygen is strongly diminished, and where nitrate reduction is maximal (Jorissen et al. 1995, 1998; Fontanier et al. 2005; Koho et al. 2008; Mojtahid et al. 2010).

Our comparison of tests sizes at the beginning and end of the experiment indicates that *M. barleeanus* is able to calcify under the three imposed laboratory conditions. Oxygen concentration had no significant effect on the growth rate of *M. barleeanus,* so that calcification does not seem to be affected by such conditions. The fact that no reproduction was observed may be explained, like with *A. tepida*, by the fact that no food was added during the experiment. However, since reproduction of this species has never been observed before in laboratory conditions, it is difficult to draw firm conclusions on this subject.

Surprisingly, *Bulimina marginata* was strongly affected by all our laboratory conditions, with low survival rates, both under oxic and hypoxic conditions. This result is surprising because, unlike many other deep-sea foraminiferal taxa, *B. marginata* previously showed a good adaptation to laboratory conditions (Wilson-Finelli et al. 1998; Havach et al. 2001; Hintz et al. 2004, 2006a, b; McCorkle et al. 2008; Barras et al. 2009, 2010; Filipsson et al. 2010).

Nevertheless, our results show no significant difference between the three different treatments. The survival rates of *B. marginata* are not significantly lower in the aquaria with long-term hypoxia alternating with short anoxic periods, with or without nitrate. This result is in agreement with the literature. *Bulimina marginata* is known to be able to live under hypoxic and anoxic conditions in the natural environment (Bernhard and Sen Gupta 1999). In many older studies, *B. marginata* has been considered as a good marker of low oxygen conditions (e.g. Phleger and Soutar 1973; van der Zwaan and Jorissen 1991; Sen Gupta and Machain-Castillo 1993; Bernhard and Sen Gupta 1999). However, some experimental studies have shown that oxygen-depleted conditions can have a negative effect on this species. In a 24-day experiment, the survival rate of *B. marginata* in nitrogen-flushed waters was significantly lower than in the oxic controls (Bernhard and Alve 1996). According to Ernst et al. (2005), *B. marginata* was hardly influenced by hypoxia, but higher abundances where observed in oxygenated microcosms. Alve and Bernhard (1995) noticed that specimens migrated to the sediment surface and occurred on polychaete tubes when the oxygen condition became lower than 0.2 mL L^{-1}. They interpreted this migration as behavior to avoid decreasing oxygen contents within the sediment. Conversely, Moodley et al. (1998) performed laboratory experiments to test the viability of foraminifera in anoxia; *Bulimina marginata* was still found alive (Rose Bengal stained) after a maximum of 21 days of anoxic incubation.

In our study, growth of juveniles of *B. marginata* was recorded in oxic as well as hypoxic conditions. Although the percentage of individuals that added new chambers is lower in oxic conditions (15 % versus 25 % and 30 % in the hypoxic aquaria), the number of added chambers was higher than in hypoxic conditions. Unfortunately, our dataset is too limited to draw any firm conclusion.

Bulimina marginata is able to reproduce under oxic conditions, producing juveniles which have grown and calcified. Such reproductions were observed at T1, T3 and T4 with 80, 37 and 142 counted juveniles, respectively. In experiences described by Barras et al. (2009), in which low quality food was added, adult specimens of *B. marginata* produced an average of 30 juveniles per reproduction event. If the number of juveniles per reproduction had been the same in our experiments, we can estimate the numbers of adults that reproduced (3 at T1, 1 at T3 and 4 to 5 at T4). The apparent absence of reproduction in our hypoxic aquaria contrasts with results obtained by Alve and Bernhard (1995). During a 4-week mesocosm experiment with low oxygen conditions, these authors observed Rose Bengal stained juveniles of *B. marginata,* suggesting that reproduction took place.

Two main facts stand out in the data obtained for *B. marginata*: (1) the survival rates of *B. marginata* are low in all tested conditions, and (2) juveniles were observed under oxic conditions but not under hypoxic conditions. According to the literature, *Bulimina marginata* is relatively easy to keep alive, and reproduction can be obtained under laboratory condition by feeding them with fresh or frozen diatoms or green algae (Hintz et al. 2004, 2006a, b; McCorkle et al. 2008; Barras et al. 2009; Filipsson et al. 2010). However, no reproduction was observed when no additional food was added to the cultures (C. Barras, Pers. Com). To explain our surprising observations, three contrasted hypotheses can be envisaged.

Hypothesis 1

In our experiment, high mortality rates of *B. marginata* could be explained by unfavorable conditions in our aquaria due to the lack of fresh organic matter. The only source of foraminiferal food was the organic matter existing in the fine sediment added. We chose not to add food in order to avoid complex geochemical processes in the vials. However, reproduction occured under oxic conditions. Boltovskoy and Wright (1976) discussed how reproduction could be a stress response or a response to favorable conditions. Our results may lead to contradictory hypothesis. Either food is lacking and could explain high mortality rate and reproduction events as a response to stress condition; or food is not a limiting factor and reproduction occurred in response to favorable condition. In this case, another parameter may be responsible for the death of *B. marginata*.

Hypothesis 2

The absence of juvenile specimens in the hypoxic/anoxic treatments suggests that hypoxic conditions have a negative impact on reproduction. We could hypothesize that foraminifera did not reproduce under hypoxia, which could lead a decrease of the metabolic rate. Such a diminished metabolic rate could be sufficient for cell growth and calcification, but insufficient for reproduction.

Hypothesis 3

Asexual reproduction of foraminifera usually results in the death of the parent individual, creating the presence of empty adult tests (Lee et al. 1991; Barras et al. 2009). For *Bulimina marginata*, Barras et al. (2009) observed that all adults died after reproduction. In our study, reproduction occurred in oxic conditions explaining a large part of the empty adult tests, thereby contributing in a significant way to the low survival rates observed for *B. marginata*. Very similar survival rates were observed under oxic and hypoxic conditions. We may therefore wonder if the low survival rates in hypoxic conditions could also be partly explained by reproduction events. The only evidence we have that no reproduction took place in the hypoxic aquaria is the absence of juveniles. However, Alve and Goldstein (2003) demonstrated by experimental approaches that reproduction of foraminifera produces propagules (uncalcified chamber smaller than 63 μm) which are able to stay in the sediment for at least 4 months without any growth and calcification. This phenomenon may also have occurred in our experiment. Specimens of *B. marginata* may have reproduced and, perhaps, the produced propagules never calcified. At the end of the experiment, sediment and water containing foraminifera were sieved over a 63 μm mesh and >63 μm fractions were observed. With this protocol, it is very well possible that propagules <63 μm were present at the end of the hypoxic experiments.

A very important consequence of this hypothesis would be that hypoxia does not have an impact on reproduction but does have a large impact on propagule calcification, which did not take place. Our results, obtained for the growth rates (chamber addition and calcification) of *B. marginata,* support this idea (lower calcification rate under hypoxia than under oxic conditions). The ultimate consequence of this hypothesis would be that the successive disappearance of foraminiferal species after long term hypoxia/anoxia (as observed in Mediterranean sapropels, e.g., Jorissen 1999) would not be explained by adult mortality, but rather because juveniles are unable to calcify. Unfortunately, our present data set does not allow us to make firm conclusions at this moment; additional experiments are needed to confirm or refute this hypothesis.

10.5.2 Metabolism in Hypoxic Conditions

Aerobic respiration rates that were measured in this study for *Melonis barleenus* (0.32 nL O_2 cell^{-1} h^{-1} (±0.26)) are in the same range as those published by Geslin et al. (2011), who reported a minimum respiration rate value of 0.09 nL O_2 cell^{-1} h^{-1} (±0.02) for *Rectuvigerina phlegri* and a maximum value of 5.27 nL cell^{-1} h^{-1} (±0.52) for *Ammonia beccarii.* As it has been shown previously, respiration rates of foraminifera vary in function of cell size following the equation $R = 3.98\ 10^{-3}\ BioVol^{0.88}$ (with respiration rate (R) expressed in nL O_2 h^{-1} and biovolume (BioVol) in μm^{-3}), for the 17 species studied (Geslin et al. 2011). The same authors showed that benthic foraminifers have lower oxygen respiration rates than other groups of meiofauna, even when standardized for biovolume. This suggests that foraminifera have a relatively low oxygen demand, which may explain why their aerobic metabolism is less affected by low oxygen contents (our hypoxic treatments). The lower respiration rate may reflect a lower metabolic rate; which could in turn be explained by the low degree of activity of the foraminifera. A relatively low metabolic rate is advantageous for organisms exposed to environmental stress (Theede et al. 1969). In this specific case of oxygen deficiency, a low metabolic rate may help the foraminifer to preserve its energy pool during short anoxic periods.

10.5.3 Intracellular Nitrate Content

10.5.3.1 Average Data of Intracellular Nitrate Content

The intracellular nitrate contents recorded in the present study are in the same order of magnitude as the data published by Piña-Ochoa et al. (2010a). The average of intracellular nitrate contents calculated for the numerous measurements made for the three studied species varies between 16 to 61 pmol NO_3^- per cell (Table 10.4) whereas measurements reported by Piña-Ochoa et al. (2010a) range between 3 to 40 pmol NO_3^- cell^{-1}. However, the nitrate concentrations of the three studied species are still very low (max. value 462 pmol NO_3^- per cell) compared to the much higher values reported by Piña-Ochoa et al. (2010a) for *Cyclammina cancellata*

(45,500 pmol NO_3^- per cell) or *Globobulimina turgida* (18,000 pmol NO_3^- per cell). The high concentrations of intracellular nitrates allow these two species to respire nitrates under anoxia for at least 3 months (Piña-Ochoa et al. 2010a).

Ammonia tepida was reported as non nitrate-storage species by Piña-Ochoa et al. (2010a). In our case, a higher number of specimens was measured (n=29) and, although 13 specimens showed no intracellular nitrate, the other had intracellular nitrate with a maximum value of 114 pmol NO_3^- per cell (23.3 mM of NO_3^-). Knowing that the maximum concentration of nitrate in the surrounding sea water is 50 μM, these new data show that also *A. tepida* is able to store nitrate in their cells. However, these results do not imply that this species is also actively denitrifying. In Piña-Ochoa et al. (2010a), no positive denitrification rates were obtained for *Ammonia tepida*. However, it is probable that this negative result is explained by the fact that the specimens used by Piña-Ochoa et al. (2010a) did not contain any intracellular nitrate. In order to clarify this point, denitrification rates should be measured using specimens with elevated intracellular nitrate contents.

The specimens of *M. barleeanus* measured in the present study show higher nitrate contents than those reported by Piña-Ochoa et al. (2010a). However, the nitrate contents of *M. barleeanus* are still very low compared to other species such as *Uvigerina elongatastriata* (5,400 pmol NO_3^- per cell), which occupies the same intermediate infaunal microhabitat as *M. barleeanus* (e.g., Koho et al. 2008; Mojtahid et al. 2010). Piña-Ochoa et al. (2010a) suggested that *M. barleeanus* would not be able to denitrify because of the insufficient amount of nitrate available in the cell. In the light of our new data, this conclusion should be reconsidered. However, in order to ascertain that *M. barleeanus* is indeed able to respire nitrate under anoxia, it is absolutely necessary to measure denitrification rates. The fact that roughly all analyzed specimens of *M. barleeanus* collected on the continental Margin off Portugal contain intracellular nitrate (n=55, our study) whereas specimens collected in the Rhone delta (n=2, Piña-Ochoa et al. 2010a) did not (Table 10.4), suggests that environmental parameters influence the nitrate content of this species.

10.5.3.2 Impact of the Three Tested Conditions on Intracellular Nitrate Contents

The intra-specific variability of intracellular nitrate concentrations within a population is high. The same observation was made by Koho et al. (2011) who reported that the nitrate concentration in single individuals of living *G. turgida* ranged from 0 to 32,541 pmol N per cell, corresponding to average concentrations of 3,929±4,590 and 8,999±9,023 pmol N per cell (SEM higher than the mean) in two replicate cores. A similarly high intra-population variability in nitrate concentrations has also been noted by Piña-Ochoa et al. (2010a, b) and Bernhard et al. (2011, 2012). Because of this high variability, it is not easy to follow the story of the intracellular nitrate in individual foraminifera. Nevertheless, our data show a single statistically significant trend.

Intracellular nitrate contents do not show significant changes with time for *Ammonia tepida* and *Melonis barleeanus* (Fig. 10.7). If these species are able to denitrify, we could expect that they should use nitrate at least during the short periods of anoxia. This should lead to a decrease of their intracellular nitrate concentration in the hypoxic aquaria without nitrate during anoxia. Furthermore, if these 2 species would shift to nitrate respiration in case of anoxia, in the hypoxic aquaria without nitrate they should die after the exhaustion of their pools of intracellular nitrate. Therefore, all our data suggest that *A. tepida* and *M. barleeanus* did not denitrify but nevertheless survived hypoxia and short periods of anoxia.

The only significant trend is the decrease of nitrate contents in *B. marginata* with time in hypoxic conditions without nitrate which could be explained by the fact that this taxon has indeed used nitrate for foraminiferal denitrification during the short periods of anoxia. It is possible that such a decreasing trend in nitrate content has not been observed in the hypoxic aquaria with nitrate because the foraminifera have recharged their intracellular nitrate pool after each of the anoxic periods. Our data therefore suggest the capacity of *B. marginata* to denitrify. However, the significant trend could be discussed because of the high variability of nitrate content. The previous suggestions should be resolved by additional studies with measurements of denitrification rates.

10.5.4 Discussion Regarding Experimental Methodologies

10.5.4.1 Survival Determination

Working with foraminifera in oxygen-depleted environments can be problematic because of the determination of living foraminifera during the experiment. Observation of Rose Bengal stained cytoplasm inside the test is not sufficient to judge vitality when incubations are performed under oxygen-depleted conditions or for short periods of time, because cytoplasm can be retained in dead specimens for weeks to months (e.g. Bernhard 1988; Hannah and Rogerson 1997). Accurate methods to distinguish living foraminifera are needed. We chose to use fluorogenic probes which are non fluorescent compounds that produce a fluorescent product after modification by intracellular esterases that are only active in living individuals. Two different fluorogenic probes may be used: CTG (Cell Tracker Green) and FDA (Bernhard et al. 1995, 2006). CTG is a fluorogenic probe producing fluorescence which can be fixed with formalin (Bernhard et al. 2006; Pucci et al. 2009). Fixed samples may be observed after the fixation (weeks to months) using an epifluorescence binocular. On the opposite, FDA, another fluorogenic probe, cannot be fixed with formalin (Bernhard, 2000) so that samples treated with FDA have to be observed few hours after the incubation. Consequently, CTG is more practical when the amount of samples to observe is too high to be treated within a few hours. The inconvenience of this probe is its prohibitive price. In our case, we observed the fluorescence of FDA incubated specimens a few hours after the end of the experiment to assess vitality. The same protocol was used by Piña-Ochoa et al. (2010b) and Koho et al. (2011).

10.5.4.2 Growth Observations

In our study, two methodologies were used to determine growth of the incubated specimens: introduction of marked-shell specimens using calcein (Bernhard et al. 2004) and automatic measurements of the foraminiferal shell size before and after the experiment. The advantage of the calcein method is to be able to detect growth of specimens even if growth is very small (e.g. one chamber) and to determinate the number of additional chambers (unmarked) calcified during the incubation. Determination of growth is also possible with precise size measurements when considering each specimen separately at the beginning and at the end of the experiment. However, when considering the average size of a pool of individuals, which is the case in our study (practically impossible to have 1 specimen per vial and a statistically significant numbers of individuals per condition), a slight growth of a limited number of individuals of this pool might be hidden in the standard error of the average size of all specimens.

10.5.4.3 Reproduction

In many experiments, reproduction events are identified thanks to the production of juveniles (e.g. Le Cadre and Debenay 2006; Barras et al. 2009). The first step of the foraminiferal life is a single uncalcified chamber called a "propagule" (Alve and Goldstein 2003). These progagules may calcify chambers in order to become juvenile foraminifera, but they also may remain as uncalcified propagules for many weeks to months (Alve and Goldstein 2003). Consequently, reproduction events may occur without the production of calcified juveniles. To prove accurately the occurrence of reproduction events during experiments, it is therefore necessary to observe the entire samples (no sieving) and particularly the small fraction (< 63 μm) with a microscope in order to avoid loss of propagules.

10.6 Conclusion

Ammonia tepida and *Melonis barleeanus* show a similar response to hypoxic conditions alternating with short periods of anoxia. We have recorded similarly very high survival rates under oxic and hypoxic conditions as well as similar growth rates. No reproduction was observed, which may be due to the lack of added food. Therefore, additional experiments are needed to demonstrate their ability to reproduce under low oxygen conditions.

Bulimina marginata shows a very different response. The survival rates are much lower, but there was no significant difference in survival between the oxic and hypoxic/anoxic treatments.

It is possible that the high mortality is due to unfavorable experimental conditions (e.g. lack of food). Alternatively, the observation of juveniles in the oxic treatment suggests that the high mortality rate may be at least partially due to reproduction. The absence of calcified juveniles in the hypoxic/anoxic treatments can result from an absence of reproduction (in which case hypoxia would inhibit reproduction), but could also be explained by the incapacity of the juveniles to calcify, forcing them to stay at a propagule stage.

Intracellular nitrate contents suggest that *B. marginata* may be able to denitrify during short periods of anoxia.

This study allows us to propose future experimental work with interesting species such as *M. barleeanus* which was not previously used for laboratory experiments.

Acknowledgements We thank the crews of Côte de la Manche (PECH cruise) and of the Pelagia (PACEMAKER cruise) for good collaboration during the cruise. We thank the essential work of H. Howa for organizing the cruise and for sampling sediment in the Bay of Biscay. We tanks grateful N. Risgaard-Petersen for his help to analyze intracellular nitrate contents in the foraminifera. This work was founded by the French national program (EC2CO from CNRS).

References

Aller RC (1994) Bioturbation and remineralization of sedimentary organic matter: effects of redox oscillation. Chem Geol 114(3–4):331–345

Alve E, Bernhard JM (1995) Vertical migratory response of benthic foraminifera to controlled oxygen concentrations in an experimental mesocosm. Mar Ecol Prog Ser 116:137–151

Alve E, Goldstein ST (2003) Propagule transport as a key method of dispersal in benthic foraminifera (Protista). Limnol Oceanogr 48(6):2163–2170

Alve E, Murray JW (2001) Temporal variability in vertical distributions of live (stained) intertidal foraminifera, southern England. J Foraminiferal Res 31(1):12–24

Barmawidjaja DM, Jorissen FJ, Puskaric S, van der Zwaan GJ (1992) Microhabitat selection by benthic foraminifera in the northern Adriatic Sea. J Foraminiferal Res 22:297–317

Barnett PRO, Watson J, Connely D (1984) A multiple corer for taking virtually undisturbed sample from shelf, bathyal and abyssal sediments. Oceanol Acta 7:399–408

Barras C, Geslin E, Duplessy JC, Jorissen F (2009) Optimisation of laboratory conditions to obtain reproduction and growth of the deep-sea benthic foraminifer *Bulimina marginata*. J Foraminiferal Res 39(3):155–165

Barras C, Duplessy JC, Geslin E, Michel E, Jorissen F (2010) Calibration of $\delta^{18}O$ of laboratory-cultured deep-sea benthic foraminiferal shells in function of temperature. Biogeosciences 7(1):1349–1356

Bernhard JM (1988) Postmortem vital staining in benthic Foraminifera: duration and importance in population and distributional studies. J Foraminiferal Res 18:143–146

Bernhard JM (2000) Distinguishing live from dead foraminifera: methods review and proper applications. Micropaleontology 46:38–46

Bernhard JM, Alve E (1996) Survival, ATP pool, and ultrastructural characterization of benthic foraminiferafrom Drammensfjord (Norway): response to anoxia. Mar Micropaleontol 28(1):5–17

Bernhard JM, Sen Gupta BK (1999) Foraminifera of oxygen-depleted environments. In: Sen Gupta BK (ed) Modern Foraminifera. Kluwer Academic, Dordrecht

Bernhard JM, Newkirk SG, Bowser SS (1995) Towards a non-terminal viability assay for Foraminiferan Protists. J Eukaryot Microbiol 42(4):357–367

Bernhard JM, Blanks JK, Hintz CJ, Chandler GT (2004) Use of the fluorescent calcite marker calcein to label foraminiferal tests. J Foraminiferal Res 34:96–101

Bernhard JM, Ostermann DR, Williams D, Blanks JK (2006) Comparison of two methods to identify live benthic foraminifera: a test between Rose Bengal and CellTracker Green with implications for stable isotope paleoreconstructions. Paleoceanography 21(4): art. no. PA4210

Bernhard J M, Edgcomb VP, Casciotti KL, McIlvin MR, Beaudoin DJ (2011) Denitrification likely catalyzed by endobionts in an allogromiid foraminifer. ISME J 6:951–960. doi:10.1038/ismej.2011.171

Bernhard JM, Casciotti KL, McIlvin MR et al (2012) Potential importance of physiologically diverse benthic foraminifera in sedimentary nitrate storage and respiration. J Geophys Res 117(G3), G03002. doi:10.1029/2012JG001949

Bollmann J, Quinn P, Vela M et al (2004) Automated particle analysis: calcareous microfossils. In: Image analysis, sediments and paleoenvironments. Kluwer, Dordrecht

Boltovskoy E, Wright R (1976) Recent foraminifera. W. Junk, The Hague, 515 p

Bouchet VMP, Debenay J-P, Sauriau P-G, Radford-Knoery J, Soletchnik P (2007) Effects of short-term environmental disturbances on living benthic foraminifera during the Pacific oyster summer mortality in the Marennes-Oléron Bay (France). Mar Environ Res 64:358–383

Bouchet V, Sauriau P-G, Debenay J-P, Mermillod-Blondin F (2009) Influence of the mode of macrofauna-mediated bioturbation on the vertical distribution of living benthic foraminifera: first insight from axial tomodensitometry. J Exp Mar Biol Ecol 371:20–33

Bradshaw JS (1957) Laboratory studies on the rate of growth of the foraminifer "*Streblus beccarii* (Linné) var. *tepida* (Cushman)". J Paleontol 31:1138–1147

Bradshaw JS (1961) Laboratory experiments on the ecology of foraminifera. Contrib Cushman Found Foramin Res 12:87–106

Braman RS, Hendrix SA (1989) Nanogram nitrite and nitrate determination in environmental and biological materials by vanadium (III) reduction with chemiluminescence detection. Anal Chem 61:2715–2718

Buzas MA (1977) Vertical distribution of foraminifera in the Indian River, Florida. J Foraminiferal Res 7(3):234–237

Caralp HM (1989) Abundance of *Bulimina exilis* and *Melonis barleeanum*: relationship to the quality of marina organic matter. Geo-Mar Lett 9:37–43

Chambers JM (1992) In: Chambers JM, Hastie TJ (eds) Statistical models in S, Chapman & Hall London, available at: http://www.lavoisier.fr/livre/notice.asp?id=OKRW3SA2OKOOWW. Accessed 12 Mar 2013

Corliss BH (1991) Morphology and microhabitat preferences of benthic foraminifera from the northwest Atlantic Ocean. Mar Micropaleontol 17:195–236

De Nooijer LJ, Toyofuku T, Kitazato H (2009) Foraminifera promote calcification by elevating their intracellular pH. Proc Natl Acad Sci USA 106(36):15374–15378. doi:10.1073/pnas.0904306106

De Rijk S, Troelstra SR, Rohling EJ (1999) Benthic foraminiferal distribution in the Mediterranean Sea. J Foraminiferal Res 29:93–103

De Rijk S, Jorissen FJ, Rohling EJ, Troelstra SR (2000) Organic flux control on bathymetric zonation of Mediterranean benthic foraminifera. Mar Micropaleontol 40:151–166

Debenay J-P, Zhang J, Beneteau E et al (1998) *Ammonia beccarii* and *Ammonia tepida* (Foraminifera): morphofunctional arguments for their distinction. Mar Micropaleontol 34:235–244

Debenay J-P, Guillou J-J, Redois F, Geslin E (2000) Distribution trends of foraminiferal assemblages in paralic environments: a base for using foraminifera as bioindicators. In: Martin RE (ed) Environmental micropaleontology: the application of microfossils to environmental geology. Kluwer, New York

Diaz RJ, Rosenberg R (1995) Marine benthic hypoxia: a review of its ecological effects and the behavioural responses of benthic macrofauna. Oceanogr Mar Biol 33:245–303

Diaz R, Rosenberg R (2008) Spreading dead zones and consequences for marine ecosystems. Science 321(5891):926–929

Dissard D, Nehrke G, Reichart GJ, Bijma J (2010a) Impact of seawater pCO(2) on calcification and Mg/Ca and Sr/Ca ratios in benthic foraminifera calcite: results from culturing experiments with *Ammonia tepida*. Biogeosciences 7(1):81–93

Dissard D, Nehrke G, Reichart GJ, Bijma J (2010b) The impact of salinity on the Mg/Ca and Sr/Ca ratio in the benthic foraminifera *Ammonia tepida*: results from culture experiments. Geochim Cosmochim Acta 74(3):928–940. doi:10.1016/j.gca.2009.10.040

Diz P, Barras C, Geslin E et al (2012) Incorporation of Mg and Sr and oxygen and carbon stable isotope fractionation in cultured *Ammonia tepida*. Mar Micropaleontol 92–93:16–28

Dupuy C, Rossignol L, Geslin E, Pascal P-Y (2010) *Ammonia tepida*: a hard-shelled foraminifera predator of mudflat meio-macrofaunal metazoan. J Foraminiferal Res 40(4):305–312

Ellis BFS, Messina AR (1940) Catalogue of foraminifera. Special Publication. American Natural Museum, New York

Ernst S, Bours R, Duijnstee I, van der Zwaan BD (2005) Experimental effects of an organic matter pulse and oxygen depletion on a benthic foraminiferal shelf community. J Foraminiferal Res 35(3):177–197

Fenchel T (2012) Anaerobic eukaryotes. In: Altenbach AV, Bernhard JM, Seckbach J (eds) Anoxia: evidence for Eukaryote survival and paleontological strategies. Springer, Dordrecht

Filipsson HL, Bernhard JM, Lincoln SA, McCorkle DC (2010) A culture-based calibration of benthic foraminiferal paleotemperature proxies: $d^{18}O$ and Mg/Ca results. Biogeosciences 7:1335–1347

Fontanier C, Jorissen FJ, Licari L et al (2002) Live benthic foraminiferal faunas from the Bay of Biscay: faunal density, composition, and microhabitats. Deep Sea Res Part I 49:751–785

Fontanier C, Jorissen FJ, Chaillou G et al (2005) Live foraminiferal faunas from a 2800 m deep lower canyon station from the Bay of Biscay: faunal response to focusing of refractory organic matter. Deep Sea Res Part I 52(7):1189–1227

Fontanier C, Jorissen FJ, Lansard B et al (2008) Live foraminifera from the open slope between Grand Rhône and Petit Rhône Canyons (Gulf of Lions, NW Mediterranean). Deep Sea Res Part I 55(11):1532–1553

Frankel L (1975) Pseudopodia of surface and subsurface dwelling *Miliammina fusca* (Brady). J Foraminiferal Res 5:211–217

Geslin E, Debenay J-P, Lesourd M (1998) Abnormal textures in the wall of deformed tests of *Ammonia* (Hyaline foraminifer). J Foraminiferal Res 28(2):148–156

Geslin E, Heinz P, Hemleben C, Jorissen FJ (2004) Migratory response of deep-sea benthic foraminifera to variable oxygen conditions: laboratory investigations. Mar Micropaleontol 53:227–243

Geslin E, Risgaard-Petersen N, Lombard F et al (2011) Respiration rates of benthic foraminifera using oxygen microsensors. J Exp Mar Biol Ecol 396:108–114

Glud RN (2008) Oxygen dynamics of marine sediments. Mar Biol Res 4:243–289

Goldstein ST, Corliss BH (1994) Deposit feeding selected deep-sea and shallow-water benthic foraminifera. Deep Sea Res Part I 41:229–241

Goldstein ST, Moodley L (1993) Gametogenesis and the life cycle of the foraminifer *Ammonia beccarii* (Linne) forma *tepida* (Cushman). J Foraminiferal Res 23:213–220

Hannah F, Rogerson A (1997) The temporal and spatial distribution of foraminiferans in marine benthic sediments of the Clyde Sea area, Scotland. Estuar Coast Shelf Sci 44(3):377–383

Havach SM, Chandler GT, Wilson-Finelli A, Shaw TJ (2001) Experimental determination of trace element partition coefficients in cultured benthic foraminifera. Geochim Cosmochim Acta 65:1277–1283

Hayward BW, Holzmann M, Grenfell H, Pawlowski P, Triggs C (2004) Morphological distinction of molecular types in *Ammonia*—towards a taxonomic revision of the world's most commonly misidentifed foraminifera. Mar Micropaleontol 50:237–271

Heinz P, Geslin E (2012) Ecological and biological response of benthic foraminifera under oxygen-depleted conditions: evidence from laboratory approaches. In: Altenbach AV, Bernhard JM, Seckbach J (eds) Anoxia: evidence for Eukaryote Survival and paleontological Strategies. Springer, Dordrecht

Hess S, Jorissen F (2009) Distribution patterns of living benthic foraminifera from Cap Breton Canyon, Bay of Biscay: faunal response to sediment instability. Deep Sea Res Part I 56:1555–1578

Hess S, Jorissen FJ, Venet V, Abu-Zied R (2005) Benthic foraminiferal recovery after recent turbidite deposition in Cap Breton canyon, Bay of Biscay. J Foraminiferal Res 35:114–129

Hintz CJ, Chandler GT, Bernhard JM et al (2004) A physicochemically constrained seawater culturing system for production of benthic foraminifera. Limnol Oceanogr Methods 2:160–170

Hintz CJ, Shaw TJ, Chandler GT et al (2006a) Trace/minor element:calcium ratios in cultured benthic foraminifera. Part I: Inter-species and inter-individual variability. Geochim Cosmochim Acta 70:1952–1963

Hintz CJ, Shaw TJ, Bernhard JM et al (2006b) Trace/minor element:calcium ratios in cultured benthic foraminifera. Part II: Ontogenetic variation. Geochim Cosmochim Acta 70:1964–1976

Høgslund S, Revsbech NP, Cedhagen T, Nielsen LP, Gallardo VA (2008) Denitrification, nitrate turnover, and aerobic respiration of benthic foraminiferans in the oxygen minimum zone off Chile. J Exp Mar Biol Ecol 359:85–91

Jannink NT, Zachariasse WJ, van der Zwaan GJ (1998) Living (Rose Bengal stained) benthic foraminifera from the Pakistan continental margin (northern Arabian Sea). Deep Sea Res Part I 45:1483–1513

Jørgensen BB (2005) Oxygen distribution and bioirrigation in Arctic fjord sediment (Svalbard, Barents Sea). Mar Ecol Prog Ser 292:85–95

Jørgensen BB, Revsbech NP (1989) Oxygen-uptake, bacterial distribution, and carbon-nitrogen-sulfur cycling in sediments from the Baltic Sea North-Sea transition. Ophelia Suppl 31(1):29–49

Jorissen FJ (1988) Benthic foraminifera from the Adriatic Sea: principles of phenotypic variation. Utrecht Micropaleontol Bull 37:174

Jorissen FJ (1999) Benthic foraminiferal microhabitats. In: Sen Gupta BK (ed) Foraminifera. Kluwer, Dordrecht

Jorissen FJ, de Stigter HC, Widmark JGV (1995) A conceptual model explaining benthic foraminiferal microhabitats. Mar Micropaleontol 26(1–4):3–15

Jorissen FJ, Wittling I, Peypouquet JP, Rabouille C, Relexans JC (1998) Live benthic foraminiferal faunas off Cap Blanc, NW Africa: community structure and microhabitats. Deep Sea Res Part I 45:2157–2188

Josefson AB, Widbom B (1988) Differential response of benthic macrofauna and meiofauna to hypoxia in the Gullmar Fjord basin. Mar Biol 100(1):31–40

Justic D, Rabalais NN, Turner RE (2003) Simulated responses of the Gulf of Mexico hypoxia to variations in climate and anthropogenic nutrient loading. J Mar Syst 42:115–126

Kitazato H (1989) Vertical distribution of benthic foraminifera within sediments (Preliminary Report). Benthos Results Bull Jpn Assoc Benthol 35–36:41–51

Kitazato H (1994) Diversity and characteristics of benthic foraminiferal microhabitats in four marine environments around Japan. Mar Micropaleontol 24:29–41

Koho KA, Piña-Ochoa E (2012) Benthic foraminifera: inhabitants of low-oxygen environments. In: Altenbach AV, Bernhard JM, Seckbach J (eds) Anoxia: evidence for eukaryote survival and paleontological strategies. Springer, Dordrecht

Koho KA, Kouwenhoven TJ, de Stigter HC, van der Zwaan GJ (2007) Benthic foraminifera in the Nazaré Canyon, Portuguese continental margin: sedimentary environments and disturbance. Mar Micropaleontol 66(1):27–51

Koho KA, Langezaal AM, van Lith YA, Duijnstee IAP, van der Zwaan GJ (2008) The influence of a simulated diatom bloom on deep-sea benthic foraminifera and the activity of bacteria. A mesocosm study. Deep Sea Res Part I 55:696–719

Koho KA, Piña-Ochoa E, Geslin E, Risgaard-Petersen N (2011) Survival and nitrate uptake mechanisms of foraminifers (*Globobulimina turgida*): laboratory experiments. FEMS Microbiol Ecol 75:273–283

Langezaal AM, Jannink NT, Pierson ES, van der Zwaan GJ (2005) Foraminiferal selectivity towards bacteria: an experimental approach using a cell-permeant stain. J Exp Mar Bio Ecol 312:137–170

Le Cadre V, Debenay JP (2006) Morphological and cytological responses of ammonia (foraminifera) to copper contamination: implication for the use of foraminifera as bioindicators of pollution. Environ Pollut 143(2):304–317

Le Cadre V, Debenay J-P, Lesourd M (2003) Low pH effects on *Ammonia beccarii* test deformation: implications for using test deformations as a pollution indicator. J Foraminiferal Res 33(1):1–9

Lee JJ, Faber WW, Anderson OR, Pawlowski J (1991) Life cycles of foraminifera. In: Lee JJ, Anderson J (eds) Biology of foraminifera. Academic, London

McCorkle DC, Bernhard JM, Hintz CJ et al (2008) The carbon and oxygen stable isotopic composition of cultured benthic foraminifera. In: Austin WEN, James RH (eds) Biogeochemical controls on palaeoceanographic environmental proxies. Geological Society, London

Middelburg J, Levin A (2009) Coastal hypoxia and sediment biogeochemistry. Biogeosciences 6:1273–1293

Mojtahid M (2007) Les foraminifères benthiques: bio-indicateurs d'eutrophisation naturelle et anthropique en milieu marin franc. Ph.D. dissertation, University of Angers, Angers, 389 p

Mojtahid M, Jorissen F, Durrieu J et al (2006) Benthic foraminifera as bio-indicators of drill cutting disposal in tropical east Atlantic outer shelf environments. Mar Micropaleontol 61:58–75

Mojtahid M, Jorissen F, Pearson TH (2008) Comparison of benthic foraminiferal and macrofaunal responses to organic pollution in the Firth of Clyde (Scotland). Mar Pollut Bull 56:42–76

Mojtahid M, Griveaud C, Fontanier C, Anschutz P, Jorissen FJ (2010) Live benthic foraminiferal faunas along a bathymetrical transect (140–4800 m) in the Bay of Biscay (NE Atlantic). Rev Micropaleontol 53(3):139–162

Moodley L, Hess C (1992) Tolerance of infaunal benthic foraminifera for low and high oxygen concentrations. Biol Bull 183(1):94–98

Moodley L, Van der Zwaan GJ, Herman PMJ, Kempers L, Van Breugel P (1997) Differential response of benthic meiofauna to anoxia with special reference to Foraminifera (Protista: Sarcodina). Mar Ecol Prog Ser 158:151–163

Moodley L, van der Zwaan GJ, Rutten GMW, Boom RCE, Kempers AJ (1998) Subsurface activity of benthic foraminifera in relation to porewater oxygen content: laboratory experiments. Mar Micropaleontol 34:91–106

Moodley L, Boschker HTS, Middelburg JJ et al (2000) Ecological significance of benthic foraminifera: ^{13}C labeling experiments. Mar Ecol Prog Ser 202:289–295

Morigi C, Geslin E (2009) Quantification of benthic foraminiferal abundance. In: Danovaro R (ed) Methods for the study of deep-sea sediments, their functioning and biodiversity (from viruses to megafauna). CRC, Boca Raton

Movellan A, Schiebel R, Zubkov MV, Smyth A, Howa H (2012) Protein biomass quantification of unbroken individual foraminifers using nano-spectrophotometry. Biogeosciences 9:3613–3623

Murray JW (2006) Ecology and applications of benthic foraminifera. Cambridge University Press, Cambridge

Nardelli MP, Jorissen F, Pusceddu A et al (2010) Living benthic foraminiferal assemblages along a latitudinal transect at 1000m depth off the Portuguese margin. Micropaleontology 56:323–344

Nelder JA, Wedderburn RWM (1972) Generalized linear models. J R Stat Soc A 135(3):370–384

Oren A (2012) Diversity of anaerobic prokaryotes and eukaryotes breaking long-established dogmas. In: Altenbach AV, Bernhard JM, Seckbach J (eds) Anoxia: evidence for eukaryote survival and paleontological strategies. Springer, Dordrecht

Pascal P-Y, Dupuy C, Richard P, Niquil N (2008) Bacterivory in the common foraminifer Ammonia tepida: isotope tracer experiment and the controlling factors. J Exp Mar Bio Ecol 359:55–61

Phipps M, Jorissen FJ, Pusceddu A, Bianchelli S, De Stigter H (2012) Live benthic foraminiferal faunas along a bathymetrical transect (282–4987 M) on the Portuguese Margin (ne Atlantic). J Foraminiferal Res 42(1):66–81

Phleger FB, Soutar A (1973) Production of benthic foraminifera in three east Pacific oxygen minima. Micropaleontology 19:110–115

Piña-Ochoa E, Høgslund S, Geslin E et al (2010a) Widespread occurrence of nitrate storage and denitrification among foraminifera and gromiids. Proc Natl Acad Sci USA 107:1148–1153

Piña-Ochoa E, Koho K, Geslin E, Risgaard-Petersen N (2010b) Survival and life strategy of foraminifer, *Globobulimina turgida*, through nitrate storage and denitrification: laboratory experiments. Mar Ecol Prog Ser 417:39–49

Poag CW (1978) Paired foraminiferal ecophenotypes in gulf coast estuaries: ecological and paleoecological implications. Trans Gulf Coast Assoc Geol Soc 28:395–420

Pucci F, Geslin E, Barras C et al (2009) Survival of benthic foraminifera under hypoxic conditions: results of an experimental study using the cell tracker green method. Mar Pollut Bull 59:336–351

R Development Core Team (2011) R: a language and environment for statistical computing (version 2.14.0). R Foundation for Statistical Computing, Vienna. http://www.R-project.org/

Revsbech NP (1989) An oxygen microsensor with a guard cathode. Limnol Oceanogr 34:474–478

Revsbech NP, Madsen B, Jørgensen BB (1986) Oxygen production and consumption in sediments determined at high spatial resolution by computer simulation of oxygen microelectrode data. Limnol Oceanogr 31:293–304

Risgaard-Petersen N, Langezaal AM, Ingvardsen S et al (2006) Evidence for complete denitrification in a benthic foraminifer. Nature 443:93–96

Schmidt S, Amiard JC, Dupas B, Bradshaw JS (1957) Laboratory studies on the rate of growth of the foraminifer, "*Streblus beccarii* (Linné) var. *tepida* (Cushman)". J Paleontol 31:1138–1147

Schmiedl G, De Bovée F, Buscail R et al (2000) Trophic control of benthic foraminiferal abundance and microhabitat in the bathyal Gulf of Lions, western Mediterranean Sea. Mar Micropaleontol 40:167–188

Schnitker D (1974) Ecotypic variation in *Ammonia beccarii* (Linné). J Foraminiferal Res 4(4):217–223

Sen Gupta BK, Machain-Castillo ML (1993) Benthic foraminifera in oxygen-poor habitats. Mar Micropaleontol 20:183–201

Stouff V, Lesourd M, Debenay J-P (1999a) Laboratory observations on asexual reproduction (schizogony) and ontogeny of *Ammonia tepida* with comments on the life cycle. J Foraminiferal Res 29:75–84

Stouff V, Geslin E, Debenay J-P, Lesourd M (1999b) Origin of morphological abnormalities in *Ammonia* (foraminifera): studies in laboratory and natural environments. J Foraminiferal Res 29:152–170

Theede H, Ponat A, Hiroki K, Schlieper C (1969) Studies on the resistance of marine bottom invertebrates to oxygen-deficiency and hydrogen sulfide. Mar Biol 2:325–337

Treude T (2012) Biogeochemical reactions in marine sediments underlying anoxic water bodies. In: Altenbach AV, Bernhard JM, Seckbach J (eds) Anoxia: evidence for Eukaryote Survival and paleontological Strategies. Springer, Dordrecht

Van der Zwaan GJ, Jorissen FJ (1991) Biofacial patterns in river-induced shelf anoxia. In: Tyson RV, Pearson TH (eds) Modern and ancient continental shelf anoxia, vol 58. Geological Society Special Publication, pp 65–82

Walton WR, Sloan BJ (1990) The genus *Ammonia* Brünnich, 1792: its geographic distribution and morphologic variability. J Foraminiferal Res 20(2):128–156

Wenzhöfer F, Glud RN (2004) Small-scale spatial and temporal variability in coastal benthic O2 dynamics: effects of fauna activity. Limnol Oceanogr 49(5):1471–1481

Wilson-Finelli A, Chandler GT, Spero HJ (1998) Stable isotope behavior in paleoceanographically important benthic foraminifera: results from microcosm culture experiments. J Foraminiferal Res 28:312–320

Chapter 11
Living Foraminifera in a Brazilian Subtropical Coastal Environment (Flamengo Inlet, Ubatuba, São Paulo State–Brazil)

André Rosch Rodrigues, Teresa Lima Díaz, and Vivian Helena Pellizari

Abstract The aim of the present study is determine living (Rose Bengal stained) foraminifera fauna in the Flamengo Inlet, a subtropical inlet placed in Ubatuba a city from São Paulo State in Brazil. The inlet was sampled during the 2010 austral summer (March): 34 surface sediment samples were collected with a Van Veen grab sampler. Faunal analysis followed standard procedures, where a fixed volume of 50 cm^3 of sediment was washed through sieves with 0.50 and 0.062 mm openings and the living plus dead fauna were picked out.

The living foraminifera fauna of the study area among all the samples has a total of 32,002 specimens and is composed mainly by calcareous taxa. The main species found are *Buliminella eligantissima*, *Ammonia beccarii*, *Brizalina striatula* and *Nonionella atlantica*, all of which are calcareous and very frequent in coastal areas. The Flamengo Inlet has a typical foraminiferal fauna to subtropical coastal environments, with high values of diversity, evenness, abundance and richness revealing that the inlet is not strongly affected by severe anthropogenic impact even though the stations near some marinas, at the inner portion known as "Saco do Ribeira," showed the lowest abundance and richness values.

Keywords Non-impacted assemblages • South Atlantic foraminiferal assemblages • Subtropical coastal environment

---

A.R. Rodrigues (✉) • T.L. Díaz • V.H. Pellizari
Department of Biological Oceanography, Oceanographic Institute–University of São Paulo, São Paulo, SP 05508-900, Brazil
e-mail: andrerr@usp.br

H. Kitazato and J.M. Bernhard (eds.), *Approaches to Study Living Foraminifera: Collection, Maintenance and Experimentation*, Environmental Science and Engineering, DOI 10.1007/978-4-431-54388-6_11, © Springer Japan 2014

11.1 Introduction

All over the world, the coastal areas are most intensely affected by changes caused by anthropogenic influence (Eichler et al. 2012) and some of these changes directly influence the benthic communities and, among them, the foraminifera assemblages (e.g. Alve 1995; Cearreta et al. 2000).

A frequent problem in ecological and oceanographic surveys that try to use foraminifers as environmental proxies for most of coastal areas is that there are no data available from pre-impact times to use as reference conditions (Bouchet et al. 2012). The study of coastal systems, which may have some pristine regions, can provide fundamental information helpful to marine conservation and environmental protection. Furthermore, analysis of coastal systems could provide background data in surveys of impacted ecosystems.

In the context of the environmental variability of the long Brazilian coastline (~7,500 km), which has several different ecosystems, such as estuaries, mangroves forests, salt marshes, lagoons and beaches, the foraminiferal record in the sediment has not been extensively or intensively studied (Eichler et al. 2007). There are regions in the Brazilian coast that seems to still have natural and unaffected environments, which could yield valuable information.

The Flamengo Inlet (Fig. 11.1) is a subtropical inlet situated on the northern coast São Paulo State—Brazil, at the city of Ubatuba (23°29′42″S–23°31′30″S and 45°05′W–45°07′30″W). The inlet is open to the south and has a width of 2.3 km at its entrance. The inner portions of the inlet extending from the coastline to 5 m deep, especially the area known as "Saco do Ribeira," is characterized by weak low hydrodynamic energy and muddy sediment (Mahiques 1992). The outer part lies between 5 and 15 m deep and is strongly influenced by currents and waves from the open sea.

This inlet does not have any significant freshwater discharge besides some periodic punctual fluvial inputs, which are dependent on the local rainfall patterns. The tide currents that penetrate in this inlet make this environment relatively preserved from anthropogenic impact. The exception is the area with restricted water circulation of "Saco do Ribeira" with several marinas and hundreds of small- and medium-sized boats and constant human activity (Sanches 1992).

The aim of the present study is to determine the recently living foraminifera fauna (Rose Bengal stained specimens) in the Flamengo Inlet, a Brazilian subtropical inlet, sampled during the austral summer of 2010 and if there is any different foraminiferal assemblage to its inner portion named "Saco do Ribeira." The tide currents that penetrate in this inlet and its water circulation make this environment relatively preserved from anthropogenic impact systems but the area of "Saco do Ribeira" could be potentially impacted by human activities because of the presence of a large number of boats and many marinas at this area.

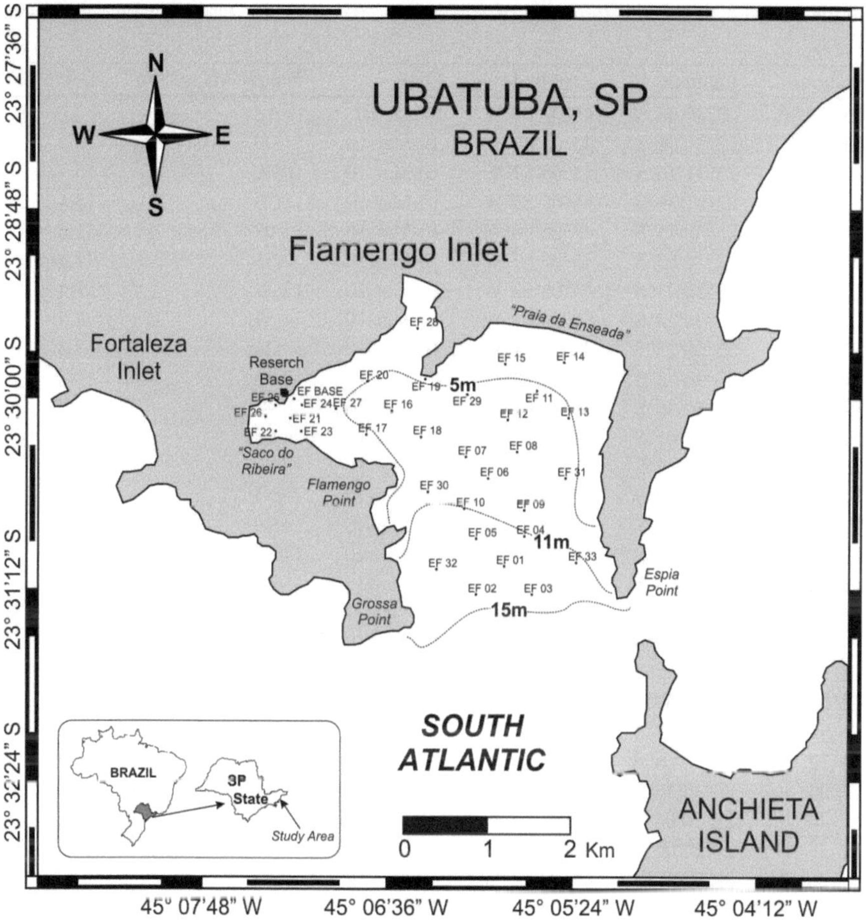

Fig. 11.1 Flamengo Inlet and the 34 stations sampled in March of 2010 (austral summer)

11.2 Materials and Methods

11.2.1 Sample Collection

The studied area was sampled during late austral summer (March) of 2010, when 34 surface sediment samples of benthic foraminifera were collected with a Van Veen grab sampler in the 2.0–13.7 m depth range (Table 11.1 and Fig. 11.1).

Table 11.1 Position and information of the 34 stations sampled at Flamengo Inlet (Ubatuba, Brazil)

Stations	Latitude	Longitude	Date	Time (GMT −3:00)	Depth (m)
EF BASE	23°30′04″S	45°07′07″W	9-Mar-10	17:18	2.0
EF 01	23°31′06″S	45°05′46″W	10-Mar-10	11:00	12.2
EF 02	23°31′16″S	45°05′57″W	10-Mar-10	10:40	13.7
EF 03	23°31′16″S	45°05′35″W	10-Mar-10	11:20	13.4
EF 04	23°30′55″S	45°05′38″W	10-Mar-10	12:02	11.6
EF 05	23°30′55″S	45°05′57″W	10-Mar-10	9:55	11.6
EF 06	23°30′33″S	45°05′52″W	10-Mar-10	12:35	9.4
EF 07	23°30′25″S	45°06′01″W	10-Mar-10	14:59	8.5
EF 08	23°30′23″S	45°05′41″W	10-Mar-10	13:28	8.5
EF 09	23°30′45″S	45°05′38″W	10-Mar-10	12:15	10.7
EF 10	23°30′44″S	45°06′01″W	10-Mar-10	9:42	10.7
EF 11	23°30′01″S	45°05′33″W	10-Mar-10	14:05	7.0
EF 12	23°30′12″S	45°05′45″W	10-Mar-10	13:40	7.6
EF 13	23°30′11″S	45°05′21″W	10-Mar-10	13:54	7.6
EF 14	23°29′50″S	45°05′23″W	10-Mar-10	14:20	5.5
EF 15	23°29′51″S	45°05′46″W	10-Mar-10	14:33	5.8
EF 16	23°30′08″S	45°06′29″W	10-Mar-10	15:40	7.3
EF 17	23°30′17″S	45°06′39″W	10-Mar-10	9:03	6.7
EF 18	23°30′18″S	45°06′18″W	10-Mar-10	15:12	7.9
EF 19	23°29′57″S	45°06′16″W	10-Mar-10	15:25	7.0
EF 20	23°29′57″S	45°06′39″W	10-Mar-10	16:14	6.7
EF 21	23°30′11″S	45°07′08″W	11-Mar-10	10:28	4.5
EF 22	23°30′16″S	45°07′14″W	11-Mar-10	9:58	3.5
EF 23	23°30′16″S	45°07′04″W	11-Mar-10	10:43	3.2
EF 24	23°30′06″S	45°07′04″W	11-Mar-10	10:58	3.8
EF 25	23°30′06″S	45°07′14″W	11-Mar-10	10:13	2.5
EF 26	23°30′11″S	45°07′18″W	11-Mar-10	9:05	3.2
EF 27	23°30′08″S	45°06′51″W	10-Mar-10	8:40	5.9
EF 28	23°29′38″S	45°06′19″W	10-Mar-10	15:54	5.8
EF 29	23°30′02″S	45°06′00″W	10-Mar-10	14:46	7.0
EF 30	23°30′38″S	45°06′15″W	10-Mar-10	9:25	9.8
EF 31	23°30′33″S	45°05′22″W	10-Mar-10	13:15	9.4
EF 32	23°31′07″S	45°06′12″W	10-Mar-10	10:10	13.1
EF 33	23°31′04″S	45°05′18″W	11-Mar-10	11:43	11.9

11.2.2 *Environmental Factors*

Some environmental factors were measured, such as depth, temperature, salinity, concentration of dissolved oxygen (DO) and pH. The content of total organic matter (TOM), Chlorophyll *a* and phaeopigments in sediment and the grain size of each sample were also determined.

The determination of bottom temperature and pH were measured with a pHmeter (MP 120, Mettler Toledo), the concentration of DO was measured with a dissolved oxygen meter (MO/28, Mettler Toledo) and bottom-water salinities measured with a handheld refractometer (ATAGO).

Sediment from all stations were collected with a plastic corer and kept frozen (−20 °C) and dark. Chlorophyll *a* and phaeopigments were extracted from each sample with 10 ml 100 % acetone together with 0.07 g $MgCO_3$ for 24 h in the dark at 4 °C. Absorbance was read at optical densities of 750, 665, and 430 nm in a spectrophotometer, before and after acidification with 1 N HCl, according to Plante-Cuny (1978). The spectrophotometric technique yields reliable results (Daemen 1986; Cahoon 1999; Skowronski et al. 2009). Results are given in mg/m^2.

Granulometric analysis was carried out using about 30 g of each sample. Hydrogen peroxide (H_2O_2 10 %) was added to this material until the TOM was completely eliminated. The remaining material was dried at 60 °C, weighed, and then washed through a 0.062 mm sieve. The coarser fraction was dried, weighed, and sieved through a series of 12 sieves from 4.0 to 0.062 mm.

The fraction finer than 0.062 mm was put in suspension in 1,000 mL of distilled water to which 1 g of sodium diphosphate ($Na_4P_2O_7$) was added to avoid flocculation of mineral clay particles. Suguio's method (1973) was then used for pipette analysis of the silt and clay fraction.

The distribution of the size fractions was computed using the TURBO (Geological Institute of the University of São Paulo) computer program.

11.2.3 Foraminifera Laboratory Procedures

After collection of the uppermost layer of the sediment (about 2 cm), the samples were stored in a mixture of 4 % formaldehyde solution and 1 g of Rose Bengal (protein stain) in 1 L of distilled water.

Faunal analysis followed standard procedures, where a fixed volume of 50 cm^3 of sediment was washed through sieves with 0.250 and 0.062 mm openings. The residue was split into subsamples using a dry microsplitter (model from Green Geological Services) of approximately 500 specimens (non-stained and stained foraminifers) or more than 100 stained specimens were counted. All specimens were counted when the number of foraminifers in a sample was less than 500. Species identifications and counting of dry specimens were done under an optical microscope (magnification of 68×). The identifications were based on the following publications that illustrate foraminiferal fauna from South Atlantic and coastal ecosystems: Brady (1884), Boltovskoy et al. (1980), Jones (1994) and Loeblich and Tappan (1988).

11.2.4 Data Analyses

The absolute and relative frequency for each species were calculated. Shannon–Wiener diversity, Simpson dominance, and Evenness were determined by PRIMER

(Clarke and Warwick 1994), using the following formulas in which pi is the proportion of the species in the sample and ln pi is the natural log of pi:

1. Shannon Wiener diversity H' = sum (pi ln pi);
2. Simpson dominance C = sum (pi2); and
3. Evenness J' = H'/H' max.

Matrices were constructed using the Bray–Curtis similarity measure on log (x + 1) to normalize foraminiferal counts. Some samples with less than 20 specimens and/or less 7 species per sample were not included (EF21, EF24 and EF25). Species-abundance data were calculated for foraminiferal samples (each contributing ≥0.5 % or more to the assemblage in more than one station) and subjected to Q-mode cluster analysis to define foraminiferal assemblages. The Bray-Curtis distance was used to measure proximity between samples, and Ward's linkage method was used to arrange samples into a hierarchical dendrogram.

The foraminiferal data were coordinated by non-metric multidimensional scaling (MDS) (Clarke and Warwick 1994) to emphasize the geometrical aspects of similarity and to enable visualization of complex data in a graphical environment. Pearson correlation analyses were performed in order to further explore more the relationships between the abiotic parameters with the main species and ecological indices, considering $p < 0.05$ as the significant level.

11.3 Results

11.3.1 Environmental Parameters

11.3.1.1 Bottom Water

The bottom water data are presented in Table 11.2.

The lowest temperature measured was 23.7 °C, at station EF 05, next to the entrance of the inlet, and the highest value was 28.9 °C, at stations EF BASE and EF 24, placed inside the inner portion of the Flamengo Inlet called "Saco do Ribeira." The average temperature was 25.7 °C.

There is a clear gradient of temperature in the inlet where the warmer values are at its inner portions, like "Saco do Ribeira," and cooler bottom waters are found adjacent to the mouth of the Inlet (Fig. 11.2).

The lowest salinity found was 30.0, at station EF 22, inside the area of "Saco do Ribeira," the highest value was 40.0, at station EF 33, near the east portion of the Inlet entrance. These are extreme values not reflecting most of the stations that show low variations of salinity, with an average value of 34.2.

The lowest content of dissolved oxygen was 1.76 mL L^{-1}, at stations EF 10, next to the entrance of the inlet, the highest value was 5.48 mL L^{-1}, at station EF 11, next the "Praia da Enseada" area. The average dissolved oxygen is 4.01 mL L^{-1}. The dissolved oxygen did not show any gradient across the studied area.

Table 11.2 Abiotic factors of the bottom water from 34 stations sampled at Flamengo Inlet (Ubatuba, Brazil)

Stations	Temperature (°C)	Salinity	Dissolved oxygen (mL L^{-1})	pH
EF BASE	28.9	35.0	4.24	6.65
EF 01	24.2	34.0	4.54	7.32
EF 02	24.1	34.5	4.15	7.45
EF 03	25.0	32.0	4.46	7.24
EF 04	24.0	34.0	4.3	7.29
EF 05	23.7	35.0	4.01	7.23
EF 06	25.4	34.5	3.87	7.31
EF 07	25.5	34.5	4.3	7.25
EF 08	26.1	34.0	4.39	6.61
EF 09	24.6	34.5	4.72	7.38
EF 10	24.6	35.0	1.76	7.11
EF 11	25.7	35.0	5.48	7.08
EF 12	25.2	34.0	4.48	6.93
EF 13	25.3	34.0	4.12	6.99
EF 14	25.2	34.0	4.26	7.13
EF 15	25.7	34.5	4.43	7.11
EF 16	25.0	34.5	3.7	7.23
EF 17	25.2	32.0	4.74	6.76
EF 18	26.6	34.0	4.26	7.22
EF 19	25.8	33.5	3.86	7.18
EF 20	26.0	34.0	3.9	7.26
EF 21	27.1	33.0	4.02	6.28
EF 22	26.9	30.0	2.87	6.48
EF 23	27.2	33.0	4.16	6.76
EF 24	28.9	34.0	3.78	6.28
EF 25	28.7	34.0	3.03	6.33
EF 26	27.0	34.0	3.84	6.54
EF 27	26.5	38.0	2.78	6.41
EF 28	25.8	34.0	3.98	7.24
EF 29	25.8	34.5	4.4	7.18
EF 30	24.4	35.0	4.03	7.03
EF 31	25.5	33.0	4.03	6.47
EF 32	24.8	35.0	4.29	7.24
EF 33	24.5	40.0	3.2	7.32

The lowest pH was 6.28, at stations EF 21 and EF 24, inside the area of "Saco do Ribeira," the highest value was 7.45, at station EF 02, next the inlet's entrance. The average pH was 6.98. The highest values of pH occurred at the central portion and entrance of the Flamengo Inlet.

11.3.1.2 Granulometric Patterns and Sediment Geochemistry

The granulometric and geochemical characteristics of the stations are given in Tables 11.3 and 11.4 and Fig. 11.3. These data show that the Flamengo Inlet surface

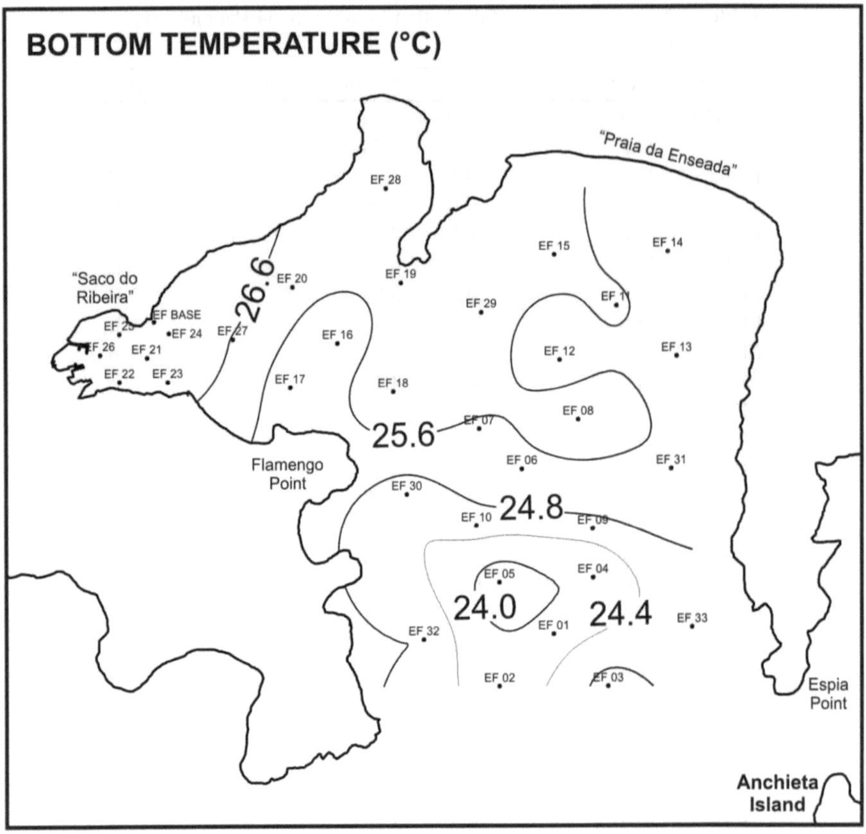

Fig. 11.2 The gradient of bottom temperature across the Flamengo Inlet measured in March of 2010 (austral summer)

sediment contains high amounts of silt and sand. Most of the stations can be classified as sandy silt or silty sand. At the "Saco do Ribeira," inner portion of the inlet, the stations show large amounts of clay. The stations EF 32 and EF 30, at the west side of the inlet's entrance, show the largest amount of sand.

The lowest surface sediment concentration of total organic matter (TOM) was 1.47 %, at station EF 10 and the highest value was 11.56 %, at station EF 01, both stations (EF 01 and EF 10) are placed near the entrance of the inlet. The average TOM is 5.02 % and the concentration of TOM does not show any gradient across the studied area.

The lowest content of chlorophyll *a* was 4.28 mg m^{-2}, at stations EF 21, and the highest content was 118.51 mg m^{-2}, at stations EF BASE, both stations are inside the "Saco do Ribeira." The average content of chlorophyll *a* is 30.69 mg m^{-2}. The lowest content of phaeopigments was 1.23 mg m^{-2}, at station EF 21, and the highest content was 163.32 mg m^{-2}, at station EF BASE, similar to the content of chlorophyll range.

Table 11.3 Gran size content and Shepard's classification of the stations sampled at Flamengo Inlet (Ubatuba, Brazil)

Station	Gravel (%)	Sand (%)	Silt (%)	Clay (%)	Shepard classification (1957)
EF BASE	14.82	72.28	8.06	4.83	Sand
EF 01	2.38	20.23	54.74	22.65	Clayey silt
EF 02	0.13	45.34	29.74	24.79	Silty sand
EF 03	0.29	73.61	13.05	13.05	Clayey sand
EF 04	1.71	12.93	74.30	11.07	Silt
EF 05	1.31	35.21	48.10	15.39	Sandy silt
EF 06	0.11	58.32	41.53	0.04	Silty sand
EF 07	0.06	63.95	35.95	0.03	Silty sand
EF 08	0.88	31.81	67.27	0.03	Sandy silt
EF 09	0.33	28.39	57.92	13.37	Sandy silt
EF 10	0.00	63.74	25.90	10.36	Silty sand
EF 11	0.00	9.15	83.15	7.70	Silt
EF 12	0.40	17.57	70.23	11.80	Sandy silt
EF 13	0.00	11.99	79.82	8.19	Silt
EF 14	0.00	7.34	83.01	9.65	Silt
EF 15	0.00	10.46	84.01	5.53	Silt
EF 16	0.58	36.65	45.93	16.84	Sandy silt
EF 17	0.08	29.50	46.95	23.47	Sandy silt
EF 18	0.00	41.48	52.13	6.40	Sandy silt
EF 19	1.12	35.84	58.26	4.78	Sandy silt
EF 20	0.00	36.54	56.78	6.68	Sandy silt
EF 21	0.00	4.52	76.82	18.66	Silt
EF 22	7.63	75.71	11.66	5.00	Sand
EF 23	0.00	16.08	52.93	30.99	Clayey silt
EF 24	0.00	14.41	51.71	33.88	Clayey silt
EF 25	0.00	6.24	55.03	38.73	Clayey silt
EF 26	0.00	7.34	60.05	32.61	Clayey silt
EF 27	0.00	38.92	38.86	22.23	Silty sand
EF 28	0.06	70.46	24.99	4.50	Silty sand
EF 29	0.34	20.40	70.83	8.43	Sandy silt
EF 30	0.30	86.61	13.08	0.00	Sand
EF 31	1.00	12.61	65.65	20.73	Clayey silt
EF 32	0.00	77.61	16.42	5.97	Sand
EF 33	0.65	34.84	48.39	16.13	Sandy silt

The average content of phaeopigments is 22.02 mg m^{-2}. The content of chlorophyll *a* and phaeopigments does not show any gradient across the Flamengo Inlet.

11.3.2 Living Foraminiferal Fauna

A total of 32,002 foraminifera were found (Table 11.5, Appendix Tables and Taxonomic list), belonging to 88 taxa, most of them identified to the level of species,

Table 11.4 Geochemistry content of the sediment sampled at the Flamengo Inlet (Ubatuba, SP)

Stations	TOM (%)	Chlorophyll *a* (mg m^{-2})	s.d.	Phaeopigments (mg m^{-2})	s.d.
EF BASE	2.12	118.51	19.13	163.32	40.71
EF 01	11.56	25.35	6.10	10.73	5.52
EF 02	8.84	31.72	12.89	32.85	23.42
EF 03	4.78	49.14	7.54	38.03	8.70
EF 04	6.60	40.54	16.38	29.20	18.95
EF 05	5.28	58.98	25.65	53.28	44.90
EF 06	3.52	39.00	2.58	25.60	6.66
EF 07	2.83	21.17	12.39	12.54	11.00
EF 08	3.49	51.76	20.81	26.16	19.07
EF 09	3.47	28.07	20.61	19.14	20.18
EF 10	1.47	14.74	1.88	7.44	0.56
EF 11	3.97	28.53	13.82	17.21	15.00
EF 12	2.77	22.72	2.55	12.71	3.40
EF 13	4.16	18.28	4.52	13.78	3.03
EF 14	4.54	18.70	17.67	11.08	13.03
EF 15	3.78	38.79	33.27	19.15	22.29
EF 16	4.73	17.24	18.80	9.76	9.74
EF 17	5.67	19.81	3.28	9.19	2.78
EF 18	2.83	48.17	16.63	22.60	15.55
EF 19	2.56	49.16	12.54	25.29	8.67
EF 20	3.76	–	–	–	–
EF 21	8.95	4.94	2.06	1.23	0.71
EF 22	3.73	27.87	6.66	12.13	5.82
EF 23	8.60	9.12	3.80	3.45	1.14
EF 24	8.64	9.26	2.65	3.15	0.66
EF 25	7.14	8.36	4.68	2.46	1.92
EF 26	7.35	9.89	3.98	2.13	1.16
EF 27	7.55	13.41	[a]	8.44	[a]
EF 28	2.45	79.63	[a]	58.12	[a]
EF 29	4.04	21.42	10.55	11.34	7.27
EF 30	2.66	18.38	6.95	5.73	2.35
EF 31	7.18	13.65	4.05	7.27	7.25
EF 32	2.06	40.30	17.67	43.39	39.18
EF 33	7.53	16.23	9.15	8.66	6.56

[a]Stations with no replicates

where most of them were calcareous (78 taxa). Others include one planktonic species and nine agglutinated species. Three stations, all from "Saco do Ribeira," an inner portion of the Flamengo Inlet (EF21, EF24 and EF25), had less than 20 individual specimens and/or not common (less than seven species per sample) species and were not used to cluster and MDS analysis. The foraminiferal abundance of the other stations ranges between 17 specimens (EF26), at the "Saco do Ribeira," to 4,634 (EF18) at the west side of the central portion of the inlet.

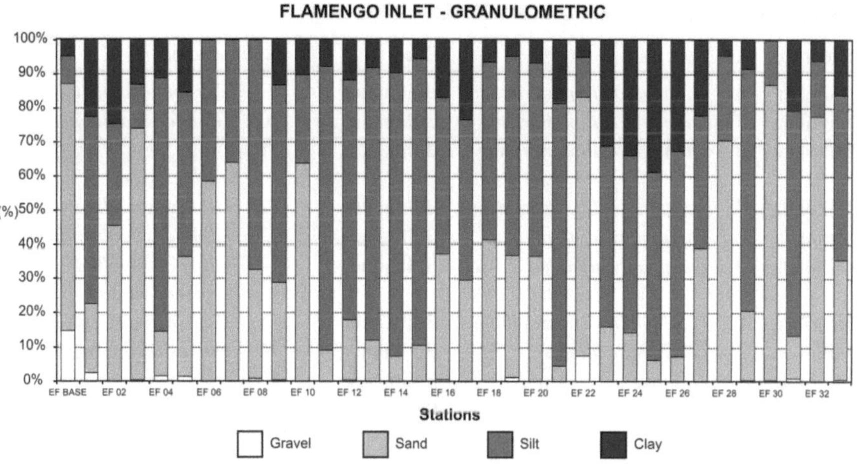

Fig. 11.3 Grain size found at the studied stations

Table 11.5 Living foraminifera taxa found at Flamengo Inlet and their orders. The most frequent taxa are in bold

Order	No.	Species	Order	No.	Species
Rotaliida	**1**	***Ammonia beccarii***	Lagenida	46	*Lagena striata*
Rotaliida	2	*Ammonia parkinsoniana*	Lagenida	47	*Oolina lineata*
Rotaliida	3	*Ammonia* sp.	Lagenida	48	*Oolina melo*
Rotaliida	4	*Cancris sagra*	Lagenida	49	*Oolina* sp.
Rotaliida	5	*Discorbis bertheloti*	Lagenida	50	*Robolus* sp.
Rotaliida	6	*Discorbis peruvianus*	Buliminida	51	*Angulogerina angulosa angulosa*
Rotaliida	7	*Discorbis williansoni*	Buliminida	52	*Bolivina* sp.
Rotaliida	8	*Elphidium advenum var. depressulum*	Buliminida	53	*Brizalina danvillensis*
Rotaliida	9	*Elphidium articulatum*	Buliminida	54	*Brizalina difformis*
Rotaliida	10	*Elphidium discoidale*	Buliminida	55	*Brizalina doniezi*
Rotaliida	11	*Elphidium excavatum*	Buliminida	56	*Brizalina lowmani*
Rotaliida	12	*Elphidium galvestonense*	Buliminida	57	*Brizalina ordinaria*
Rotaliida	13	*Elphidium gunteri*	Buliminida	58	*Brizalina* sp.
Rotaliida	14	*Elphidium* sp.	Buliminida	**59**	***Brizalina striatula***
Rotaliida	15	*Eponides repandus*	Buliminida	60	*Brizalina subaenaerensis*
Rotaliida	16	*Hanzawaia boueana*	Buliminida	61	*Brizalina translucens*
Rotaliida	**17**	***Nonionella atlantica***	Buliminida	62	*Brizalina variabilis*
Rotaliida	18	*Rosalina* sp.	Buliminida	63	*Bulimina aculeata*
Rotaliida	19	*Rosalina suezensis*	Buliminida	64	*Bulimina gibba*
Miliolida	20	*Cribromiliolinella subvalvularis*	Buliminida	65	*Bulimina marginata*
Miliolida	21	*Pyrgo narsuta*	Buliminida	66	*Bulimina patagonica*
Miliolida	22	*Pyrgo ringens*	Buliminida	67	*Bulimina* sp.
Miliolida	23	*Quinqueloculina arctica*	Buliminida	**68**	***Buliminella elegantissima***

(continued)

Table 11.5 (continued)

Order	No.	Species	Order	No.	Species
Miliolida	24	*Quinqueloculina atlantica*	Buliminida	69	*Fursenkoina pontoni*
Miliolida	25	*Quinqueloculina horrida*	Buliminida	70	*Globocassidulina crassa*
Miliolida	26	*Quinqueloculina intrincata*	Buliminida	71	*Globocassidulina laevigata*
Miliolida	27	*Quinqueloculina lamarckiana*	Buliminida	72	*Globocassidulina minuta*
Miliolida	28	*Quinqueloculina milletti*	Buliminida	73	*Globocassidulina* sp.
Miliolida	29	*Quinqueloculina patagonica*	Buliminida	74	*Globocassidulina subglobosa*
Miliolida	30	*Quinqueloculina seminulum*	Buliminida	75	*Hopkinsina pacifica*
Miliolida	31	*Quinqueloculina* sp.	Buliminida	76	*Uvigerina bifurcata*
Miliolida	32	*Quinqueloculina stalkeri*	Buliminida	77	*Uvigerina striata*
Miliolida	33	*Triloculina baldai*	Buliminida	78	*Virgulina riggi*
Miliolida	34	*Triloculina* sp.			
Miliolida	35	*Triloculina trigonula*	Globigerinida	79	*Globigerinoides* sp.
Lituolida	36	*Ammotium salsum*			
Lituolida	37	*Gaudryina exilis*	Trochamminida	80	*Trochammina inflata*
Lituolida	38	*Glomospira gordialis*	Trochamminida	81	*Trochammina ochracea*
Lituolida	39	*Haplophragmoides manillaensis*	Trochamminida	82	*Trochammina quadriloba*
Lituolida	40	*Reophax curtus*	Trochamminida	83	*Trochammina* sp.
Lituolida	41	*Reophax* sp.	Trochamminida	84	*Trochammina squamata*
Lagenida	42	*Fissurina laevigata*	Textulariida	85	*Textularia candeiana*
Lagenida	43	*Fissurina pulchella*	Textulariida	86	*Textularia earlandi*
Lagenida	44	*Lagena hispidula*	Textulariida	87	*Textularia foliacea*
Lagenida	45	*Lagena laevis*	Textulariida	88	*Textularia* sp.

The dominant species (Fig. 11.4) were the calcareous *Ammonia beccarii*, *Buliminella elegantissima*, *Ammonia parkinsoniana*, *Brizalina striatula* and *Nonionella atlantica*. Even though the genus *Ammonia* is very frequent, the species *B. elegantissima* is predominant in most of the stations, particularly those with higher values of abundance and richness (Table 11.6).

The average abundance is 941 specimens per sample and the richness values range from 2 taxa, at station EF 21, to 31, at station EF 18; the average richness is 13 species per sample. The highest diversity value was found at station EF 18, in the west side of the central portion of the inlet (2.931), while the lowest value was found at station EF 17, in the portion between the Saco do Ribeira area and the Flamengo Inlet's central portion (0.975). The average diversity is 2.043. For the same stations, the dominance index had an inverse relationship. The average dominance is 0.194.

The evenness index has its highest values found at station EF 26 (0.951), inside the "Saco do Ribeira," a sample with only 17 specimens of 9 different species, while the lowest value is at station EF 14 (0.664), in the north part of the inlet, next to the "Praia da Enseada." The average evenness is 0.818.

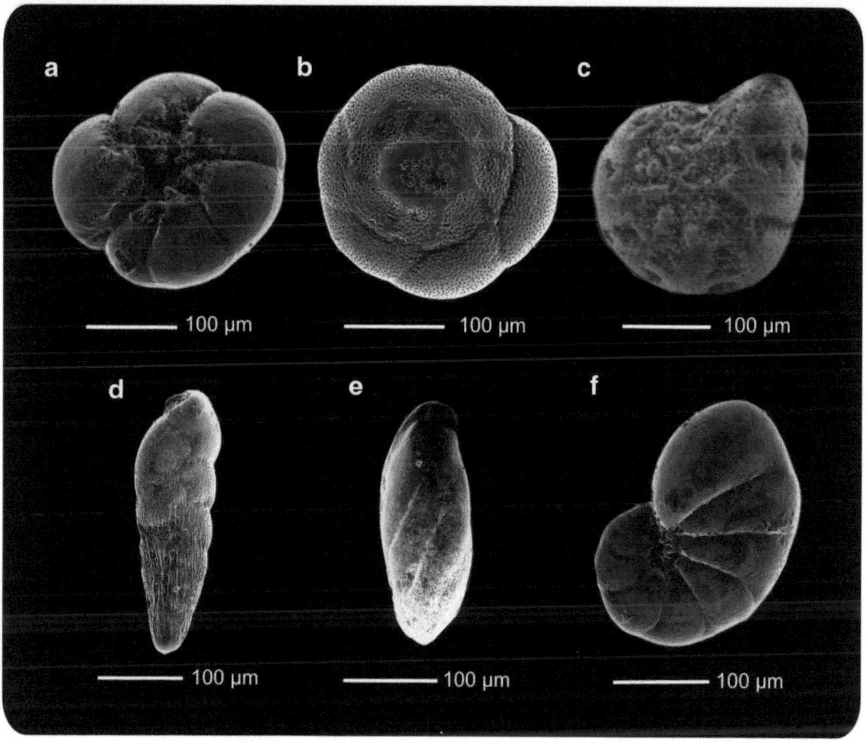

Fig. 11.4 Some of the main species found at the Flamengo Inlet: *Ammonia beccarii* (**a–b**), *Elphidium excavatum* (**c**), *Brizalina striatula* (**d**), *Buliminella elegantissima* (**e**), and *Nonionella atlantica* (**f**)

11.3.3 Numerical Analysis

The foraminiferal data matrix was made with the most representative species (each contributing 0.5 % or more to the assemblage) and subjected to the Q-Mode cluster analysis revealed clearly two main groups, one of them with two subgroups (Fig. 11.5).

Group 1 includes most of the stations, mainly those located in the entrance and central portion of the inlet, but also the station EF22 from the inner portion (Saco do Ribeira). This group can be divided into two subgroups. Group 1-A includes stations EF08, EF18, EF07, EF12, EF06, EF30, EF29, EF32 and EF19, from the central portion of the inlet, especially from the west side.

Group 1-B includes the samples EF28, EF33, EF04, EF09, EF10, EF03, EF05, EF02, EF13, EF15, EF20, EF31, EF16, EF01, EF14 and EF22, most of them located at the entrance of the inlet, at "Praia da Enseada" and at the connection between the central portion and "Saco do Ribeira." Group 2 includes the stations from the "Saco do Ribeira" EF26, EF27, EF17, EF23 and EF Base.

Table 11.6 Ecological indexes related to the living foraminifera found at Flamengo Inlet (Ubatuba, Brazil)

Stations	Abundance	Richness	Shannon–Wiener diversity	Evenness	Simpson's dominance
EF BASE	110	23	2.753	0.878	0.093
EF 01	109	10	1.769	0.768	0.236
EF 02	336	16	2.602	0.938	0.087
EF 03	1,470	18	2.265	0.784	0.185
EF 04	720	7	1.689	0.868	0.227
EF 05	575	12	2.281	0.918	0.123
EF 06	2,565	13	1.989	0.775	0.212
EF 07	2,400	22	2.368	0.766	0.167
EF 08	750	9	1.773	0.807	0.222
EF 09	1,350	17	2.363	0.834	0.136
EF 10	650	19	2.353	0.799	0.151
EF 11	540	12	2.121	0.853	0.159
EF 12	1,476	13	2.127	0.829	0.161
EF 13	690	14	2.074	0.786	0.183
EF 14	680	11	1.592	0.664	0.315
EF 15	1,560	16	2.357	0.850	0.120
EF 16	369	13	2.191	0.854	0.150
EF 17	130	4	0.975	0.703	0.459
EF 18	4,635	31	2.931	0.853	0.082
EF 19	1,935	13	1.861	0.725	0.239
EF 20	480	9	1.835	0.835	0.187
EF 21	2	2	–	–	–
EF 22	215	10	1.695	0.736	0.296
EF 23	47	8	1.602	0.771	0.271
EF 24	19	6	1.337	0.746	0.374
EF 25	15	5	1.362	0.847	0.298
EF 26	17	9	2.089	0.951	0.135
EF 27	216	9	1.607	0.731	0.279
EF 28	1,170	10	1.904	0.827	0.201
EF 29	1,980	18	2.406	0.832	0.134
EF 30	1,875	17	2.395	0.845	0.123
EF 31	350	13	2.214	0.863	0.145
EF 32	2,025	18	2.410	0.834	0.131
EF 33	540	10	2.128	0.924	0.135

The same foraminiferal data matrix used before with the overall dominant species was subjected to non-metric multidimensional scaling (MDS) analysis and showed segregation of foraminiferal assemblages similar to the cluster analysis (Fig. 11.6).

The MDS efficiency is demonstrated by the stress, which represents the necessary adjustment to represent the community in few dimensions (stress ≤ 0.2 = good ordination). The groups are distinguished by the abundance of foraminifers in each station. Group 1 shows higher values of abundance and richness of species than

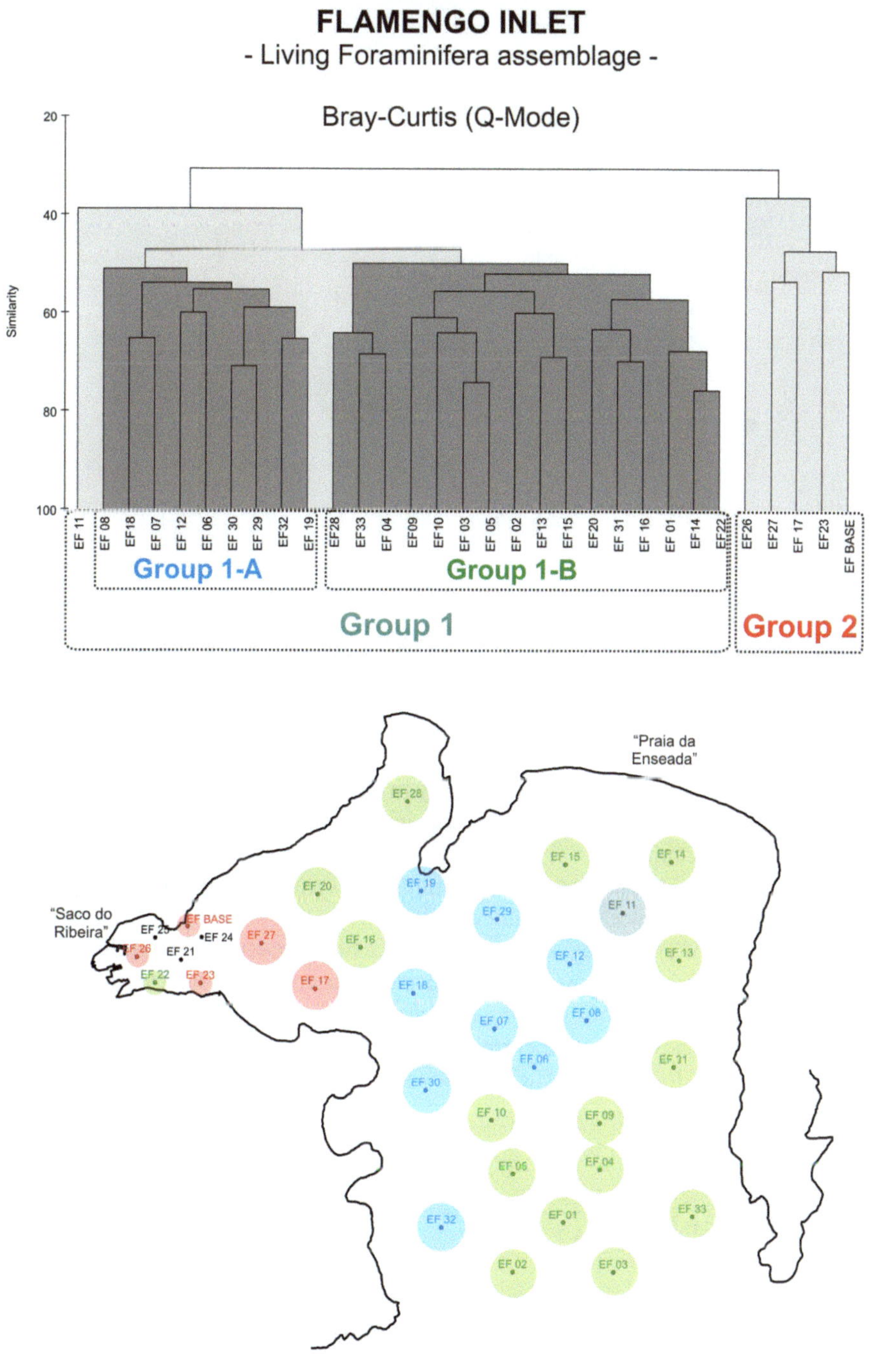

Fig. 11.5 Dendrogram classifications (Q-mode) showing the groups of stations based on faunal composition and its distribution in the Flamengo Inlet map

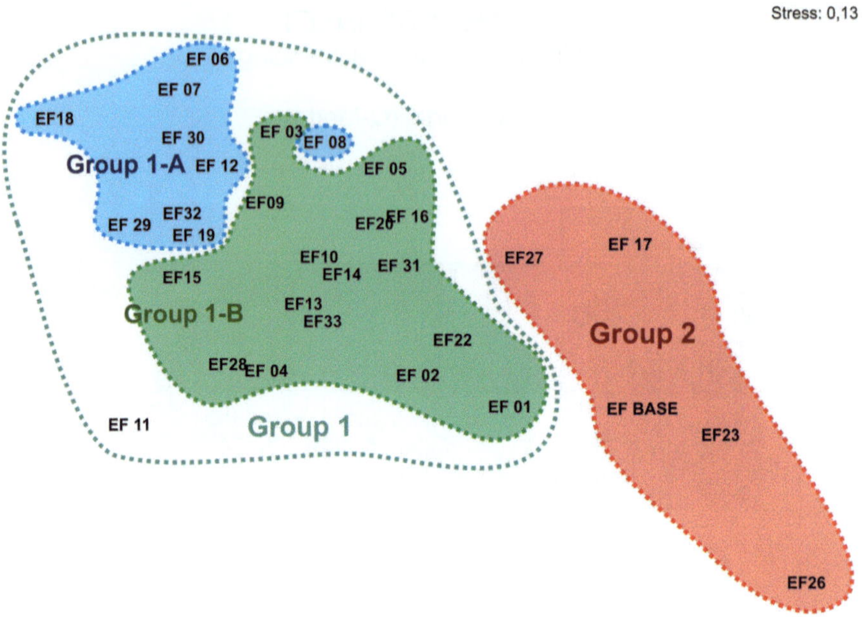

Fig. 11.6 MDS analysis with the main living species found at Flamengo Inlet

Group 2, where the species *A. beccarii* is more representative. The species *B. elegantissima* is more abundant in Group 1, mainly in Group 1-A. *N. atlantica* was common in Group 1-B, where the abundance and richness data show high values but not higher than in Group 1-A.

Table 11.7 shows results of the correlation analysis performed with the biological (foraminifera abundance) environmental data.

The content of TOM, clay and mud each has a negative correlation with diversity, abundance, richness and most abundant species. The content of sand shows positive correlation with the miliolids *Quinqueloculina patagonica* and *Q. milletti.*

The salinity shows positive correlation with evenness and negative correlation with Simpson's dominance. The pH shows positive correlation with abundance, *A. beccarii* and *B. striatula.*

11.4 Discussion

The Flamengo Inlet shows by its physical-chemical factors characteristics that it is a typical subtropical inlet. The freshwater discharge is not able to markedly change the salinity at the time of sampling. The values of dissolved oxygen show that most of the stations are normoxic (>2.0 mL L^{-1}) and pH shows slightly lower values in the inner portions, especially at "Saco do Ribeira" where the water circulation is

Table 11.7 Pearson correlation between environmental data and living foraminifera

	Depth	Temperature	Salinity	D.O.	pH	TOM	Chlorophyll *a*	Phaeopigments	Gravel	Sand	Silt	Clay	Mud
Buliminella elegantissima	0.09	−0.02	0.03	0.11	0.34	**−0.48**	0.06	−0.10	−0.25	0.21	0.04	**−0.59**	−0.17
Ammonia beccarii	0.20	−0.09	−0.13	0.12	**0.38**	**−0.48**	0.24	0.03	−0.21	0.22	−0.04	**−0.42**	−0.18
Brizalina striatula	0.33	−0.14	0.12	0.21	**0.38**	**−0.40**	0.05	−0.01	−0.23	0.26	−0.05	**−0.49**	−0.22
Nonionella atlantica	0.20	−0.02	−0.04	0.10	0.18	**−0.40**	0.12	−0.09	−0.23	0.17	0.05	**−0.52**	−0.14
Ammonia parkinsoniana	0.04	−0.06	0.06	0.10	0.33	−0.28	0.33	0.12	−0.14	0.06	0.05	−0.26	−0.05
Quinqueloculina patagonica	0.19	−0.14	0.07	0.09	0.24	**−0.36**	−0.10	−0.12	−0.17	**0.42**	−0.27	**−0.40**	**−0.38**
Quinqueloculina milletti	0.16	−0.16	0.10	0.06	0.10	−0.30	−0.04	−0.04	−0.11	**0.48**	−0.36	−0.35	**−0.45**
Elphidium galvestonense	−0.14	0.10	0.03	0.18	0.20	−0.25	−0.07	−0.16	−0.18	−0.10	0.27	**−0.36**	0.11
Pyrgo ringens	0.02	0.00	0.02	0.12	0.13	−0.34	−0.04	−0.13	−0.15	0.15	0.01	**−0.38**	−0.12
Uvigerina striata	0.01	0.02	0.05	0.16	0.21	−0.33	0.01	−0.09	−0.15	0.05	0.10	−0.33	−0.03
Elphidium articulatum	−0.11	0.01	0.08	0.24	0.25	−0.13	0.18	0.04	−0.15	−0.19	0.32	−0.23	0.20
Pyrgo narsuta	0.02	0.13	0.04	0.05	0.10	−0.18	0.09	−0.05	−0.09	0.10	−0.02	−0.19	−0.09
Quinqueloculina intrincata	−0.05	0.19	−0.02	0.07	0.12	−0.17	0.14	−0.01	−0.09	−0.03	0.10	−0.14	0.04
Elphidium discoidale	0.04	−0.12	0.01	0.13	0.18	−0.11	0.06	−0.01	−0.11	−0.18	0.24	−0.08	0.19
Quinqueloculina sp.	0.04	−0.10	−0.06	0.06	0.09	−0.28	0.03	−0.07	−0.06	0.02	0.11	−0.32	−0.01
Brizalina danvillensis	0.06	−0.03	0.02	0.02	0.19	−0.13	0.09	0.02	−0.08	0.07	0.04	−0.27	−0.06
Brizalina variabilis	0.14	−0.05	0.01	−0.03	0.17	−0.07	0.06	0.03	−0.08	0.17	−0.10	−0.18	−0.16
Brizalina sp.	0.34	−0.16	0.04	0.05	0.17	−0.24	0.04	0.09	−0.10	**0.41**	−0.35	−0.20	**−0.38**
Globocassidulina laevigata	0.25	−0.19	−0.04	0.11	0.27	−0.09	0.18	0.09	−0.10	0.22	−0.16	−0.16	−0.20
Quinqueloculina atlantica	−0.03	−0.08	0.06	0.09	0.07	−0.16	−0.16	−0.15	−0.08	0.04	0.03	−0.17	−0.03
Trochammina sp.	0.07	0.10	0.11	−0.03	0.18	−0.14	0.00	−0.09	−0.13	0.14	−0.05	−0.23	−0.12
Hanzawaia boueana	0.12	0.08	−0.07	0.04	0.17	−0.13	0.18	0.04	−0.09	0.08	−0.05	−0.07	−0.07
Discorbis williamsoni	0.14	0.08	−0.07	0.08	0.23	−0.22	0.12	−0.01	−0.13	0.29	−0.18	−0.30	−0.27
Elphidium sp.	0.24	−0.29	−0.03	0.21	0.26	−0.07	0.07	0.07	−0.10	−0.01	0.01	0.01	0.02
Fissurina laevigata	0.03	0.10	0.00	0.25	0.20	−0.33	0.14	0.02	−0.14	0.00	0.13	−0.27	0.02

(continued)

Table 11.7 (continued)

	Depth	Temperature	Salinity	D.O.	pH	TOM	Chlorophyll *a*	Phaeopigments	Gravel	Sand	Silt	Clay	Mud
Elphidium excavatum	0.05	−0.06	0.04	0.00	0.13	−0.33	0.13	0.10	−0.18	0.13	−0.02	−0.24	−0.10
Quinqueloculina arctica	0.09	−0.01	0.02	0.18	0.25	−0.22	−0.04	−0.07	−0.10	0.13	−0.04	−0.22	−0.11
Brizalina doniezi	0.09	0.16	0.04	0.16	0.09	−0.27	0.15	0.04	−0.12	0.16	−0.05	−0.26	−0.14
Ammontium salsum	**0.51**	−0.30	0.23	−0.34	0.26	0.15	−0.04	−0.01	−0.11	0.28	−0.36	0.18	−0.26
Reophax curtus	−0.02	0.19	−0.03	0.06	0.11	−0.14	0.13	−0.01	−0.07	0.02	0.03	−0.11	−0.01
Quinqueloculina lamarckiana	−0.01	0.18	−0.02	0.08	0.15	−0.19	0.09	−0.03	−0.09	0.09	−0.01	−0.19	−0.08
Brizalina translucens	0.07	0.02	0.02	0.15	0.27	−0.17	−0.02	−0.07	−0.12	0.13	−0.03	−0.22	−0.11
Virgulina riggi	−0.05	0.08	0.00	0.09	0.13	−0.31	−0.01	−0.10	−0.14	0.01	0.11	−0.27	0.00
Rare species	0.27	0.00	−0.08	0.15	**0.38**	**−0.46**	0.32	0.18	−0.13	0.34	−0.21	**−0.36**	−0.31
Shannon–Wiener diversity	0.25	0.02	0.24	0.03	0.32	−0.34	0.33	**0.37**	0.16	0.30	−0.23	−0.27	−0.31
Evenness	0.24	−0.12	**0.36**	0.07	0.17	0.12	0.16	0.23	0.06	−0.06	−0.02	0.18	0.05
Simpson's dominance	−0.27	0.06	**−0.37**	−0.02	−0.33	0.21	−0.23	−0.28	−0.07	−0.15	0.09	0.18	0.15
Abundance	0.20	−0.05	0.01	0.17	**0.41**	**−0.52**	0.16	−0.03	−0.26	0.27	−0.04	**−0.56**	−0.23
Richness	0.14	0.14	0.07	0.03	0.28	**−0.48**	0.33	0.34	0.17	**0.39**	−0.28	**−0.41**	**−0.40**

restricted and there is more intense human activity than in the rest of the inlet. The depth at this inner portion of the inlet is shallower than its entrance and the temperature is higher.

The living foraminiferal species composition shows the presence of high numbers of calcareous species frequently associated with inner shelf environments, mainly the taxa of the orders Rotaliida, Miliolida and Buliminida. The presence of agglutinated species is relatively rare, probably because of the low discharge of freshwater inside the inlet (Mahiques 1992).

The dominant species of Flamengo Inlet found in the present study are commonly found in other Brazilian coastal ecosystems (e.g. Setty 1982; Murray 1991; Bonetti 2000; Burone and Pires-Vanin 2006; Eichler et al. 2007; Eichler et al. 2012). Most of the species, such as *N. atlantica*, *B. elegantissima*, and *Q. patagonica* and the genera *Elphidium* and *Trochammina* are normally found at Ubatuba Bay (Burone and Pires-Vanin 2006) in northern of Flamengo Inlet and also in the estuarine system of Santos, São Vicente and Cubatão (Díaz et al. submitted) located in southeastern São Paulo State. Both locations are under moderate or heavy anthropogenic influence.

In ecosystems where there is any kind of anthropogenic impact or natural stress, the foraminiferal assemblages could be a useful oceanographic tool to verify these changes, not only by the presence and absence of certain species but also by the values of some ecological indices such as abundance, richness, Shannon–Wiener diversity, evenness and Simpson Dominance (Stevenson et al. 1998; Debenay et al. 1998; Eichler 2001; Eichler et al. 2003).

The existence of test deformities in coastal environments like the estuarine system of Santos, São Vicente and Cubatão and Ubatuba Inlet, both in São Paulo State (Brazil) (Geslin et al. 2002; Burone and Pires-Vanin 2006 and Díaz et al. submitted), may also indicate natural or human-induced stress. However, in the present study, deformed test were rare and without any statistical value among the whole assemblage of living foraminifers.

The groups revealed by cluster and MDS analyses were distinguished by sediment patterns and the foraminiferal fauna. The stations with high contents of silt and clay and warmer temperatures aligned with Group 2, and were mainly from the "Saco do Ribeira," inner portion of the study area, and where there were low values of abundance, richness and, consequently, diversity and evenness. The species *A. beccarii* was more frequent and the values of Simpson dominance were higher in these samples (especially sample EF17).

These aspects of Group 2 denote an area inside the Flamengo Inlet that shows an environmental condition different to the other samples in the study area, where the original ecological characteristics could be changed by the human occupation (marinas and intense boat traffic), the stations with very low abundance and richness EF21, EF24 and EF25 could be included in the Group 2.

The richness of species and the Shannon–Wiener diversity index could be used as indicators of environmental stress related to benthic foraminiferal assemblages, where the 'low' values of richness and diversity can be attributed to impacted ecosystems, as in other studies (Schafer et al. 1991; Yanko et al. 1998; Burone et al. 2006).

The prominent presence of *A. beccarii* in samples where the dominance value is high denotes a characteristic often associated with this species, as "opportunistic" in places where there are any ecological impact (e.g. Walton and Sloan 1990; Debenay et al. 2001; Burone et al. 2007; Teodoro et al. 2009).

In the present study, there is a positive correlation between the content of organic matter and the presence of *A. beccarii* or genus *Ammonia*, but the content of TOM has its highest values in "Saco do Ribeira" stations and at two entrance stations, thus, not showing a distribution gradient. However, the negative correlation of TOM with richness and abundance of foraminifers relates TOM also to Group 2.

The genus *Ammonia* is very common in coastal environments, with *A. beccarii* occurring across the Flamengo Inlet, where the foraminiferal assemblages do not show any signal of impact or natural disturbance similar to the stations associated with Group 1, where there are high amounts of sand, chlorophyll *a* and phaeopigments and the presence of Miliolids (e.g., *Quinqueloculina*) is more frequent. Lower temperatures and deeper water depths may allow a higher ecological stability to benthic systems in coastal environments than in the inner portion of the embayment (Saco do Ribeira). The western side of the entrance of the inlet seems to be under an intense influence of shelf water because it exchanges waters with the open sea, where the abiotic and biotic factor are less variable at the moment of the sampling.

Studying Ubatuba Bay, a region that is moderately affected by urban sewage located north of Flamengo Inlet, Burone et al. (2007) mention that the diversity of species cannot be a good indicator of "environmental health" in areas where there is a low abundance of foraminifera and with dominance of few species.

The ecological indices like diversity, dominance and evenness found in most of the stations at Flamengo Inlet show that these parameters could be used as indicator of "environmental health" because there is a high abundance of foraminifera in the study area, except some samples in the "Saco do Ribeira." These ecological indices show high values when compared with coastal ecosystems that are under anthropogenic impact or estuarine areas, where the natural variations could restrict the richness and diversity of species.

11.5 Summary and Conclusion

The general distribution of living foraminifera at Flamengo Inlet found in present study is summarized in Fig. 11.7, which was based on the cluster and MDS groups and ecological indices as diversity and dominance.

The Flamengo Inlet is located in the southeastern part of the Brazil and its foraminiferal fauna is composed by a diverse number of taxa mainly calcareous. The main species found during 2010 austral summer are typical from subtropical

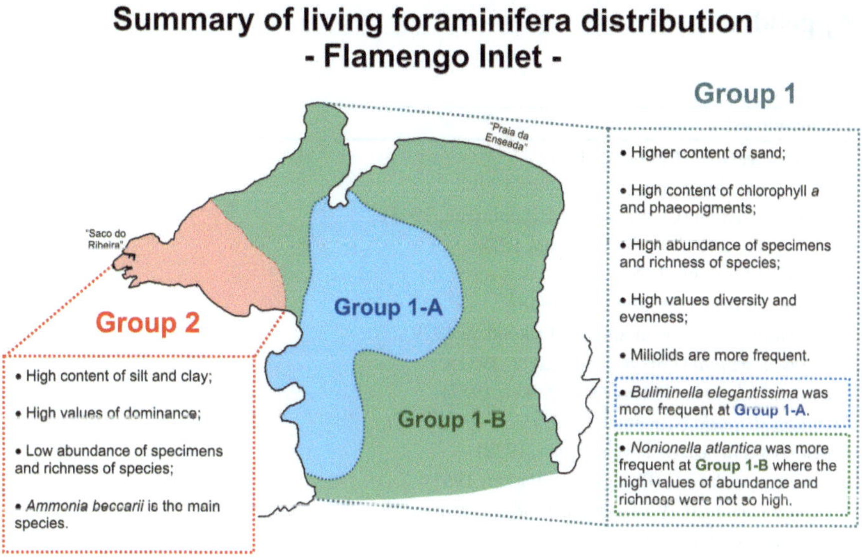

Fig. 11.7 Summary of general living foraminiferal distributions at Flamengo Inlet and its characteristics from the samples collected in March of 2010 (austral summer)

coastal environments such as species of the genera *Ammonia* and *Elphidium* and species of *N. atlantica*, *B. elegantissima*, *B. striatula* and *Q. patagonica*.

Most of the stations from the main portion and entrance of the inlet are deeper than the inner portion ("Saco do Ribeira") and show high values of diversity, evenness, abundance and richness and this area seems to be preserved and not suffering any anthropogenic impact.

Some stations at the connection of the main portion and "Saco do Ribeira," a region where there are some marinas and constant human activity, revealed lower values of abundance and richness and higher values of Simpson dominance. In general, *A. beccarii*, commonly reported in impacted areas or under environmental stress, is more frequent in the Saco do Ribiera region.

Acknowledgements The field and laboratory work was financially supported by the Fundação de Amparo à Pesquisa do Estado de São Paulo (FAPESP 2009/13666-1 and 2009/10842-3) and Oceanographic Institute of São Paulo University (IOUSP). Special thanks to the crew of the vessel "Galha Negra" and MSc. Debora do Carmo Linnhares and Mr. Diego Castillo Franco for the help in the sampling operations. We also would like to thank Drs. Joan M. Bernhard and Hiroshi Kitazato for their thorough and helpful comments on the manuscript. The principal author's Postdoctoral work was funded by a FAPESP scholarship (2009/13666-1) and he was also funded by JAMSTEC, with support to attend the "Field Workshop on Living Foraminifera in Japan".

Appendix

Taxonomic list
Ammonia beccarii (Linnaeus, 1758)
Ammonia parkinsoniana (d'Orbigny, 1839)
Ammotium salsum (Cushman and Brönnimann, 1948)
Angulogerina angulosa (Williamson, 1858)
Bolivina (d'Orbigny, 1843)
Brizalina difformis (Williamson, 1848)
Brizalina doniezi (Cushman and Wickenden, 1929)
Brizalina lowmani (Phleger and Parker, 1951)
Brizalina ordinaria (Phleger and Parker, 1952)
Brizalina striatula (Cushman, 1922)
Brizalina subaenariensis (Cushman, 1922)
Brizalina translucens (Phleger and Parker, 1951)
Brizalina variabilis (Williamson, 1858)
Bulimina aculeata (d'Orbigny, 1826)
Bulimina gibba (Fornasini, 1902)
Bulimina marginata (d'Orbigny, 1826)
Bulimina patagonica (d'Orbigny, 1839)
Buliminella elegantissima (d'Orbigny, 1839)
Cancris sagra (d'Orbigny, 1839)
Cribromiliolinella subvalvularis (Parr, 1950)
Discorbis bertheloti (d'Orbigny, 1839)
Discorbis peruvianus (d'Orbigny, 1839)
Discorbis williamsoni (Chapman and Parr, 1932)
Elphidium advenum var. *depressulum* (Cushman, 1933)
Elphidium articulatum (d'Orbigny, 1839)
Elphidium discoidale (d'Orbigny, 1839)
Elphidium excavatum (Terquem, 1875)
Elphidium galvestonense (Kornfield, 1931)
Elphidium gunteri (Cole, 1931)
Eponides repandus (Fichtel and Moll, 1798)
Fissurina laevigata (Reuss, 1850)
Fissurina pulchella (Brady, 1867)
Fursenkoina pontoni (Cushman, 1932)
Gaudryina exilis (Cushman and Brönnimann, 1948)
Globigerinoides (Cushman, 1927)
Globocassidulina crassa (d'Orbigny, 1839)
Globocassidulina laevigata (d'Orbigny, 1826)
Globocassidulina minuta (Cushman, 1933)
Globocassidulina subglobosa (Brady, 1881)
Glomospira gordialis (Jones and Parker, 1860)

(continued)

(continued)

Taxonomic list
Hanzawaia boueana (d'Orbigny, 1846)
Haplophragmoides manilaensis (Andersen, 1952)
Hopkinsina pacifica (Cushman, 1933)
Lagena hispidula (Cushman, 1913)
Lagena laevis (Montagu, 1803)
Lagena striata (d'Orbigny, 1839)
Nonionella atlantica (Cushman, 1936)
Oolina lineata (Williamson, 1848)
Oolina melo (d´Orbigny, 1839)
Pyrgo nasuta (Cushman, 1935)
Pyrgo ringens (Lamarck, 1804)
Quinqueloculina arctica (Cushman, 1933)
Quinqueloculina atlantica (Boltovskoy, 1957)
Quinqueloculina horrida (Cushman, 1947)
Quinqueloculina intricata (Terquem, 1878)
Quinqueloculina lamarckiana (d'Orbigny, 1839)
Quinqueloculina milletti (Wiesner, 1923)
Quinqueloculina patagonica (d'Orbigny, 1839)
Quinqueloculina seminulum (Linnaeus, 1758)
Quinqueloculina stalkeri (Loeblich and Tappan, 1953)
Reophax curtus (Cushman, 1920)
Robulus (de Montfort, 1808)
Rosalina suezensis (Said, 1949)
Textularia candeiana (d'Orbigny, 1839)
Textularia earlandi (Parker, 1952)
Textularia foliacea (Heron-Allen and Earland, 1915)
Triloculina baldai (Carter, 1885)
Triloculina trigonula (Lamarck, 1804)
Trochammina inflata (Montagu, 1808)
Trochammina ochracea (Williamson, 1858)
Trochammina quadriloba (Höglund, 1948)
Trochammina squamata (Jones and Parker, 1860)
Uvigerina bifurcata (d'Orbigny, 1839)
Uvigerina striata (d'Orbigny, 1839)
Virgulina riggii (Boltovskoy, 1954)

Flamengo Inlet (Ubatuba, Brazil)-living foraminifera (Part 1)									
Stations	EF BASE	EF 01	EF 02	EF 03	EF 04	EF 05	EF 06	EF 07	EF 08
Shannon–Wiener diversity	2.753	1.769	2.602	2.265	1.689	2.281	1.989	2.368	1.773
Evenness	0.878	0.768	0.938	0.784	0.868	0.918	0.775	0.766	0.807
Simpson's dominance	0.093	0.236	0.087	0.185	0.227	0.123	0.212	0.167	0.222
Total numbers	110	109	336	1470	720	575	2565	2400	750
Richness	23	10	16	18	7	12	13	22	9
FBP not identified									3.3
Ammonia beccarii	23.5	38.5	14.3	38.8	37.5	21.7	8.8	4.2	10.0
Ammonia parkinsoniana	4.9	25.6	3.6		18.8			1.0	
Ammonia sp.	2.5							1.0	
Ammontium salsum			10.7	4.1					
Angulogerina angulosa angulosa									
Bolivina sp.									
Brizalina danvillensis						4.3	7.0		
Brizalina difformis									
Brizalina doniezi	1.2							1.0	3.3
Brizalina lowmani									
Brizalina ordinaria							3.5		
Brizalina sp.		2.6		2.0		4.3		2.1	
Brizalina striatula	6.2	7.7	14.3	8.2	18.8	17.4	15.8	17.7	13.3
Brizalina subaenaerensis									
Brizalina translucens		2.6	3.6					3.1	
Brizalina variabilis	2.5		7.1	2.0			7.0		
Bulimina aculeata									
Bulimina gibba									
Bulimina marginata			3.6						
Bulimina patagonica									
Bulimina sp.									
Buliminella elegantissima	4.9	5.1	10.7	12.2	6.3	13.0	40.4	34.4	36.7
Cancris sagra									
Cribromiliolinella subvalvularis									
Discorbis bertheloti	3.7								
Discorbis peruvianus	4.9								
Discorbis williansoni		2.6		2.0			1.8	1.0	
Elphidium advenum depressum									
Elphidium articulatum			3.6		6.3				
Elphidium discoidale			3.6	2.0		4.3			
Elphidium excavatum									
Elphidium galvestonense	1.2	2.6					1.8	1.0	
Elphidium gunteri									
Elphidium sp.	1.2			2.0		8.7		1.0	
Eponides repandus	1.2								
Fissurina laevigata									
Fissurina pulchella									
Fursenkoina pontoni	1.2								
Gaudryina exilis	2.5								
Globigerinoides sp.									
Globocassidulina crassa									
Globocassidulina laevigata				4.1	6.3			2.1	

(continued)

(continued)

Flamengo Inlet (Ubatuba, Brazil)-living foraminifera (Part 1)									
Stations	EF BASE	EF 01	EF 02	EF 03	EF 04	EF 05	EF 06	EF 07	EF 08
Globocassidulina minuta			7.1	2.0					
Globocassidulina sp.									
Globocassidulina subglobosa				4.1					
Glomospira gordialis	4.9								
Hanzawaia boueana			3.6	2.0		4.3			
Haplophragmoides manillensis									
Hopkinsina pacifica				2.0		8.7			
Lagena hispidula									
Lagena laevis									
Lagena striata				2.0					
Nonionella atlantica		10.3	3.6	6.1		4.3	5.3	3.1	23.3
Oolina lineata			3.6						
Oolina melo	9.9								
Oolina sp.									
Pyrgo narsuta									3.3
Pyrgo ringens							3.5	2.1	3.3
Quinqueloculina arctica								4.2	
Quinqueloculina atlantica									
Quinqueloculina horrida									
Quinqueloculina intrincata	1.2								
Quinqueloculina lamarckiana								2.1	
Quinqueloculina milletti	3.7							2.1	
Quinqueloculina patagonica				2.0		4.3	1.8	8.3	
Quinqueloculina seminulum	3.7								
Quinqueloculina sp.					6.3			3.1	3.3
Quinqueloculina stalkeri									
Reophax curtus									
Reophax sp.									
Robolus sp.				2.0					
Rosalina sp.	1.2								
Rosalina suezensis	6.2								
Textularia candeiana									
Textularia earlandi									
Textularia foliacea			3.6						
Textularia sp.							1.8		
Triloculina baldai								1.0	
Triloculina sp.	1.2								
Triloculina trigonula	6.2								
Trochammina inflata									
Trochammina ochracea									
Trochammina quadriloba									
Trochammina sp.							1.8		
Trochammina squamata									
Uvigerina bifurcata			3.6			4.3		2.1	
Uvigerina striata		2.6							
Virgulina riggi								2.1	

Flamengo Inlet (Ubatuba, Brazil)-living foraminifera (Part 2)								
Stations	EF09	EF10	EF 11	EF 12	EF13	EF14	EF15	EF 16
Shannon–Wiener diversity	2.363	2.353	2.121	2.127	2.074	1.592	2.357	2.191
Evenness	0.834	0.799	0.853	0.829	0.786	0.664	0.850	0.854
Simpson's dominance	0.136	0.151	0.159	0.161	0.183	0.315	0.120	0.150
Total numbers	1350	650	540	1476	690	680	1560	369
Richness	17	19	12	13	14	11	16	13
FBP not identified					1.4			
Ammonia beccarii	21.7	29.2		29.3	21.7	23.5	22.1	13.6
Ammonia parkinsoniana	6.7	1.5		7.3	5.8	1.5	4.8	3.4
Ammonia sp.								
Ammontium salsum		6.2						
Angulogerina angulosa angulosa								
Bolivina sp.								1.7
Brizalina danvillensis			5.6					
Brizalina difformis	1.7				1.4	2.9		
Brizalina doniezi			2.8					
Brizalina lowmani							1.9	
Brizalina ordinaria								
Brizalina sp.		1.5	2.8			1.5		
Brizalina striatula	5.0	6.2	22.2	17.1	10.1	2.9	8.7	8.5
Brizalina subaenaerensis								
Brizalina translucens	1.7						1.0	1.7
Brizalina variabilis				2.4				
Bulimina aculeata	3.3							
Bulimina gibba		1.5						
Bulimina marginata								
Bulimina patagonica								
Bulimina sp.								
Buliminella elegantissima	25.0	21.5	27.8	17.1	33.3	50.0	16.3	30.5
Cancris sagra		1.5						
Cribromiliolinella subvalvularis								1.7
Discorbis bertheloti								
Discorbis peruvianus								
Discorbis williansoni								
Elphidium advenum depressum								
Elphidium articulatum			11.1		5.8	2.9	1.9	
Elphidium discoidale	6.7	3.1			2.9		11.5	
Elphidium excavatum		3.1	2.8		4.3		1.0	
Elphidium galvestonense			5.6		2.9		8.7	8.5
Elphidium gunteri						4.4		
Elphidium sp.	3.3					2.9		
Eponides repandus								
Fissurina laevigata			5.6				1.0	
Fissurina pulchella			2.8					
Fursenkoina pontoni					5.8			
Gaudryina exilis								
Globigerinoides sp.		4.6						
Globocassidulina crassa	1.7							
Globocassidulina laevigata				2.4				

(continued)

(continued)

Flamengo Inlet (Ubatuba, Brazil)-living foraminifera (Part 2)								
Stations	EF09	EF10	EF 11	EF 12	EF13	EF14	EF15	EF 16
Globocassidulina minuta								
Globocassidulina sp.	1.7							
Globocassidulina subglobosa								
Glomospira gordialis								
Hanzawaia boueana		1.5			1.4			
Haplophragmoides manillensis								
Hopkinsina pacifica	1.7				1.4			
Lagena hispidula	1.7							
Lagena laevis								
Lagena striata								
Nonionella atlantica	8.3	6.2		2.4	1.4	5.9	8.7	5.1
Oolina lineata								
Oolina melo								
Oolina sp.			8.3					
Pyrgo narsuta								
Pyrgo ringens				7.3				6.8
Quinqueloculina arctica	5.0							
Quinqueloculina atlantica				2.4				6.8
Quinqueloculina horrida								
Quinqueloculina intrincata		1.5					5.8	
Quinqueloculina lamarckiana								
Quinqueloculina milletti								1.7
Quinqueloculina patagonica	3.3	3.1		2.4				10.2
Quinqueloculina seminulum								
Quinqueloculina sp.		1.5		4.9				
Quinqueloculina stalkeri								
Reophax curtus								
Reophax sp.							1.0	
Robolus sp.								
Rosalina sp.								
Rosalina suezensis								
Textularia candeiana								
Textularia earlandi								
Textularia foliacea								
Textularia sp.								
Triloculina baldai								
Triloculina sp.								
Triloculina trigonula				2.4				
Trochammina inflata								
Trochammina ochracea								
Trochammina quadriloba		3.1						
Trochammina sp.								
Trochammina squamata								
Uvigerina bifurcata								
Uvigerina striata	1.7	1.5	2.8			1.5	3.8	
Virgulina riggi		1.5		2.4			1.9	

Flamengo Inlet (Ubatuba, Brazil)-living foraminifera (Part 3)									
Stations	EF 17	EF18	EF 19	EF20	EF21	EF22	EF23	EF24	EF25
Shannon–Wiener diversity	0.975	2.931	1.861	1.835	-	1.695	1.602	1.337	1.362
Evenness	0.703	0.853	0.725	0.835	-	0.736	0.771	0.746	0.847
Simpson's dominance	0.459	0.082	0.239	0.187	-	0.296	0.271	0.374	0.298
Total numbers	130	4635	1935	480	2	215	47	19	15
Richness	4	31	13	9	2	10	8	6	5
FBP not identified									33.3
Ammonia beccarii	61.5	16.5	32.6	29.7		51.2	26.2		
Ammonia parkinsoniana		3.9	4.7			8.1	7.1		
Ammonia sp.		1.0	2.3			4.7	2.4		
Ammontium salsum									
Angulogerina angulosa angulosa									
Bolivina sp.					50.0				
Brizalina danvillensis			2.3						
Brizalina difformis									
Brizalina doniezi		1.9					2.4		
Brizalina lowmani									
Brizalina ordinaria									
Brizalina sp.				1.6		4.7			
Brizalina striatula		7.8	2.3	15.6		8.1			
Brizalina subaenaerensis									
Brizalina translucens		1.0							
Brizalina variabilis									
Bulimina aculeata									
Bulimina gibba									
Bulimina marginata		1.9							
Bulimina patagonica		1.0							
Bulimina sp.				1.6					
Buliminella elegantissima	26.9	17.5	34.9	15.6		10.5	42.9		
Cancris sagra									
Cribromiliolinella subvalvularis									
Discorbis bertheloti			2.3						
Discorbis peruvianus			2.3						
Discorbis williansoni		1.9							
Elphidium advenum depressum		1.0							
Elphidium articulatum		1.0				2.3			
Elphidium discoidale									
Elphidium excavatum									
Elphidium galvestonense		1.9		10.9		1.2			
Elphidium gunteri									
Elphidium sp.	7.7								
Eponides repandus									
Fissurina laevigata	3.8	1.0	2.3						
Fissurina pulchella									
Fursenkoina pontoni						3.5			
Gaudryina exilis									
Globigerinoides sp.									
Globocassidulina crassa		1.0							
Globocassidulina laevigata									

(continued)

(continued)

Flamengo Inlet (Ubatuba, Brazil)-living foraminifera (Part 3)									
Stations	EF 17	EF18	EF 19	EF20	EF21	EF22	EF23	EF24	EF25
Globocassidulina minuta									
Globocassidulina sp.		1.0							
Globocassidulina subglobosa									
Glomospira gordialis									
Hanzawaia boueana		2.9							
Haplophragmoides manillensis		1.0							
Hopkinsina pacifica									
Lagena hispidula									
Lagena laevis									
Lagena striata									
Nonionella atlantica		3.9	4.7	18.8		5.8			
Oolina lineata									
Oolina melo									
Oolina sp.			2.3					10.5	
Pyrgo narsuta		5.8							
Pyrgo ringens		1.9							
Quinqueloculina arctica		1.0							
Quinqueloculina atlantica									
Quinqueloculina horrida									
Quinqueloculina intrincata		6.8							
Quinqueloculina lamarckiana		2.9							
Quinqueloculina milletti		1.0							
Quinqueloculina patagonica		1.9							
Quinqueloculina seminulum									
Quinqueloculina sp.			4.7						
Quinqueloculina stalkeri		1.0							
Reophax curtus		3.9			50.0		4.8	10.5	
Reophax sp.									
Robolus sp.									
Rosalina sp.									
Rosalina suezensis									
Textularia candeiana									
Textularia earlandi				1.6				10.5	
Textularia foliacea									
Textularia sp.								5.3	
Triloculina baldai									
Triloculina sp.									
Triloculina trigonula									
Trochammina inflata									6.7
Trochammina ochracea								57.9	13.3
Trochammina quadriloba									
Trochammina sp.		1.9		4.7			7.1	5.3	6.7
Trochammina squamata							7.1		40.0
Uvigerina bifurcata									
Uvigerina striata		1.9	2.3						
Virgulina riggi		1.0							

Flamengo Inlet (Ubatuba, Brazil)-living foraminifera (Part 4)								
Stations	EF26	EF27	EF28	EF 29	EF 30	EF 31	EF32	EF33
Shannon-Wiener diversity	2.089	1.607	1.904	2.406	2.395	2.214	2.410	2.128
Evenness	0.951	0.731	0.827	0.832	0.845	0.863	0.834	0.924
Simpson's dominance	0.135	0.279	0.201	0.134	0.123	0.145	0.131	0.135
Total numbers	17	216	1170	1980	1875	350	2025	540
Richness	9	9	10	18	17	13	18	10
FBP not identified							2.2	
Ammonia beccarii	11.8	30.6	34.6	27.3	16.0	14.3	22.2	16.7
Ammonia parkinsoniana	17.6	5.6	23.1	4.5	1.3	3.6	4.4	12.5
Ammonia sp.	17.6					3.6		
Ammontium salsum		2.8						8.3
Angulogerina angulosa angulosa							2.2	
Bolivina sp.								
Brizalina danvillensis				1.5		3.6		
Brizalina difformis								
Brizalina doniezi		2.8					2.2	
Brizalina lowmani								
Brizalina ordinaria								
Brizalina sp.	5.9	5.6					4.4	
Brizalina striatula				9.1	5.3	7.1	22.2	20.8
Brizalina subaenaerensis			3.8				2.2	
Brizalina translucens								
Brizalina variabilis		5.6						
Bulimina aculeata								
Bulimina gibba								
Bulimina marginata								
Bulimina patagonica								
Bulimina sp.								
Buliminella elegantissima	17.6	41.7	11.5	18.2	22.7	21.4	13.3	16.7
Cancris sagra								
Cribromiliolinella subvalvularis								
Discorbis bertheloti								
Discorbis peruvianus								
Discorbis williansoni					1.3			
Elphidium advenum depressum								
Elphidium articulatum			7.7	4.5				4.2
Elphidium discoidale								
Elphidium excavatum			3.8	1.5		3.6	2.2	
Elphidium galvestonense				3.0	2.7	3.6		
Elphidium gunteri								
Elphidium sp.		2.8		1.5				
Eponides repandus								
Fissurina laevigata				1.5			2.2	
Fissurina pulchella								
Fursenkoina pontoni							2.2	
Gaudryina exilis								
Globigerinoides sp.								
Globocassidulina crassa								

(continued)

(continued)

Flamengo Inlet (Ubatuba, Brazil)-living foraminifera (Part 4)

Stations	EF26	EF27	EF28	EF 29	EF 30	EF 31	EF32	EF33
Globocassidulina laevigata			3.8					4.2
Globocassidulina minuta								
Globocassidulina sp.								
Globocassidulina subglobosa								
Glomospira gordialis								
Hanzawaia boueana								
Haplophragmoides manillensis								
Hopkinsina pacifica						3.6		
Lagena hispidula								
Lagena laevis							2.2	
Lagena striata								
Nonionella atlantica	5.9		3.8		5.3	25.0	2.2	8.3
Oolina lineata								
Oolina melo								
Oolina sp.				1.5				
Pyrgo narsuta				1.5	4.0			4.2
Pyrgo ringens				3.0	4.0			
Quinqueloculina arctica								
Quinqueloculina atlantica				6.1	4.0			
Quinqueloculina horrida	5.9							
Quinqueloculina intrincata								
Quinqueloculina lamarckiana								
Quinqueloculina milletti			3.8	4.5	16.0	3.6	6.7	
Quinqueloculina patagonica		2.8		3.0	8.0	3.6	2.2	
Quinqueloculina seminulum								
Quinqueloculina sp.					1.3			
Quinqueloculina stalkeri								
Reophax curtus								
Reophax sp.	11.8						2.2	
Robolus sp.								
Rosalina sp			3.8					
Rosalina suezensis				1.5				
Textularia candeiana	5.9							
Textularia earlandi								
Textularia foliacea								
Textularia sp.								
Triloculina baldai								
Triloculina sp.								
Triloculina trigonula					1.3			
Trochammina inflata								
Trochammina ochracea								
Trochammina quadriloba								
Trochammina sp.					1.3	3.6		4.2
Trochammina squamata								
Uvigerina bifurcata					1.3			
Uvigerina striata				6.1	4.0		2.2	
Virgulina riggi								

References

Alve E (1995) Benthic foraminiferal responses to estuarine pollution: a review. J Foraminiferal Res 25:190–203

Boltovskoy E, Giussani G, Watanabe S, Wright R (1980) Atlas of benthic shelf foraminifera of the southwest Atlantic. Dr. W. Junk by Publishers, The Hague-Boston-London, 147

Bonetti CVDH (2000) Foraminíferos como bioindicadores do gradiente de estresse ecológico em ambientes costeiros poluídos. Estudo aplicado ao sistema estuarino de Santos-São Vicente (SP, Brasil). Ph.D. thesis, Instituto Oceanográfico, Universidade de São Paulo, São Paulo, Brazil, 229 p (Unpublished data)

Bouchet VMP, Alve E, Rygg B, Telford RJ (2012) Benthic foraminifera provide a promising tool for ecological quality assessment of marine waters. Ecol Indic 23:66–75

Brady HB (1884) Report on the foraminifera dredged by H.M.S. *Challenger* during the years 1873–1876. Zoology 9:1–814

Burone L, Pires-Vanin AMS (2006) Foraminiferal assemblages in the Ubatuba Bay, southeastern Brazilian coast. Sci Mar 70:203–217

Burone L, Venturini N, Sprechmann P, Valente P, Muniz P (2006) Foraminiferal responses to polluted sediments in the Montevideo coastal zone, Uruguay. Mar Pollut Bull 52:61–73

Burone L, Valente P, Pires-Vanin AMS, Sousa SHM, Mahiques MM, Braga ES (2007) Benthic foraminiferal variability on a monthly scale in a subtropical bay moderately affected by urban sewage. Sci Mar 71:775–792

Cahoon LB (1999) The role of benthic microalgae in neritic ecosystems: oceanography and marine biology. An Ann Rev 37:47–86

Cearreta A, Irabien MJ, Leorri E, Yusta I, Croudace IW, Cundy AB (2000) Recent anthropogenic impacts on the Bilbao estuary, northern Spain: geochemical and macrofaunal evidence. Estuar Coast Shelf Sci 50:571–592

Clarke KR, Warwick RM (1994) Change in marine communities: an approach to statistical analyses and interpretation. Natural Environment Research Council, Swindon, 144 p

Daemen EAMJ (1986) Comparison of methods for the determination of chlorophyll in estuarine sediments. J Sea Res 20:21–28

Debenay JP, Eichler BB, Duleba W, Bonetti C, Eichler PB (1998) Water stratification in coastal lagoons: its influence on foraminiferal assemblages in two Brazilian lagoons. Mar Micropaleontol 35:67–89

Debenay JP, Geslin E, Eichler BB, Duleba W, Sylvestre F, Eichler PPB (2001) Foraminiferal assemblages in a hypersaline lagoon Araruama (RJ) Brazil. J Foraminiferal Res 31:133–151

Díaz TL, Rodrigues AR, Eichler BB (in press) Distribution of foraminifera in a subtropical Brazilian estuarine system. J Foraminiferal Res 44 (prevision screen—2014)

Eichler PPB (2001) Avaliação e diagnóstico do Canal de Bertioga (São Paulo, Brasil) através da utilização de foraminíferos como indicadores ambientais. Ph.D. thesis, University of São Paulo, São Paulo, Brazil, 240 p and CD-Rom (Unpublished data)

Eichler PPB, Eichler BB, Miranda LB, Pereira ERM, Kfouri PBP, Pimenta FM, Bérgamo AL, Vilela CG (2003) Benthic foraminiferal response to variations in temperature, salinity, dissolved oxygen and organic carbon, in the Guanabara Bay, Rio de Janeiro, Brazil. Anuário do Instituto de Geociências—UFRJ 26:36–51

Eichler PPB, Eichler BB, Miranda LB, Rodrigues AR (2007) Modern foraminiferal facies in a subtropical estuarine channel, Bertioga, São Paulo, Brazil. J Foraminiferal Res 37:234–247

Eichler PPB, Eichler BB, Sen Gupta BK, Rodrigues AR (2012) Foraminifera as indicators of marine pollutant contamination on the inner continental shelf of southern Brazil. Mar Pollut Bull 64:22–30

Geslin E, Debenay JP, Duleba W, Bonetti C (2002) Morphological abnormalities of foraminiferal tests in Brazilian environments: comparison between polluted and non-polluted areas. Mar Micropaleontol 45:151–168

Jones RW (1994) The *challenger* foraminifera. Oxford University Press, Oxford, 149 p

Loeblich AR, Tappan H (1988) Foraminiferal genera and their classification, vol 2. Van Nostrand Rienhold, New York

Mahiques MM (1992) Variações temporais na sedimentação holocênica dos embaiamentos da região de Ubatuba (SP). Tese de doutorado. Universidade de São Paulo, Instituto Oceanográfico, São Paulo, Brazil, 352 p (Unpublished data)

Murray JW (1991) Ecology and paleoecology of benthic foraminifera. Longman Scientific & Technical, Harlow, 397 p

Plante-Cuny MR (1978) Pigments photosynthétiques et production primaire des fonds meubles néritiques d'une region tropicale (Nosy-Bé, Madagascar). Trav Doc ORSTM, Paris, 359 p

Sanches TM (1992) Distribuição dos foraminíferos recentes na região de Ubatuba, São Paulo. Dissertação de mestrado, Universidade de São Paulo, Instituto Oceanográfico, São Paulo, Brazil, 110 p (Unpublished data)

Schafer CT, Collins ES, Smith NJ (1991) Relationship of foraminifera and the camoebian distribution to sediments contaminated by pulp mill effluent: Saguenary Fjord, Quebec, Canada. Mar Micropaleontol 17:255–283

Setty MGAP (1982) Pollution effects monitoring with foraminifera as indices in the Thana Creek, Bombay area. Int J Environ Stud 18:205–209

Skowronski RSP, Gheller PF, Bromberg S, David CJ, Petti MAV, Corbisier TN (2009) Distribution of microphytobenthic biomass in Martel Inlet, King George Island (Antarctica). Polar Biol 32:839–851

Stevenson MR, Dias-Brito D, Stech JL, Kampel M (1998) How do cold water biota arrive in a tropical bay near Rio de Janeiro, Brazil? Cont Shelf Res 18:1595–1612

Suguio K (1973) Introdução a Sedimentologia. Edgard Blücher - EDUSP, São Paulo, Brazil, 317 p

Teodoro AC, Duleba W, Lamparelli CC (2009) Associações de foraminíferos e composição textural da região próxima ao emissário submarino de esgotos domésticos de Cigarras, Canal de São Sebastião, SP, Brasil. Pesquisas em Geociências 36:79–94

Walton WR, Sloan B (1990) The genus *Ammonia Brünnich*. 1972: its geographic distribution and morphological variability. J Foraminiferal Res 20:128–156

Yanko V, Ahmad M, Kaminski M (1998) Morphological deformities of benthic foraminiferal test in response to pollution by heavy metals: implications for pollution monitoring. J Foraminiferal Res 28:177–200

MIX
Papier aus verantwortungsvollen Quellen
Paper from responsible sources
FSC® C105338

If you have any concerns about our products,
you can contact us on
ProductSafety@springernature.com

In case Publisher is established outside the EU,
the EU authorized representative is:
Springer Nature Customer Service Center GmbH
Europaplatz 3, 69115 Heidelberg, Germany

Printed by Libri Plureos GmbH
in Hamburg, Germany